An Introduction
to Practical
Astronomy

An Introduction to Practical Astronomy

Brian Jones

CHARTWELL
BOOKS, INC.

A QUINTET BOOK

Published by Chartwell Books
A Division of Book Sales, Inc.
110 Enterprise Avenue
Secaucus, New Jersey 07094

This edition produced for sale in the U.S.A., its
territories and dependencies only.

ISBN 1–55521–728–1

This book was designed and produced by
Quintet Publishing Limited
6 Blundell Street
London N7 9BH

Creative Director: Terry Jeavons
Designer: Wayne Blades
Project Editor: Judith Simons
Picture Researcher: Brian Jones
Illustrator: Danny McBride

Typeset in Great Britain by
Central Southern Typesetters, Eastbourne
Manufactured in Singapore by Eray Scan Pte Ltd
Printed in Singapore by
Kim Hup Lee Printing Co Pte Ltd

CONTENTS

INTRODUCTION

We live on the third planet out from a yellow dwarf star we call the Sun. Appearing bright merely due to its proximity to our planet, the Sun dominates the Solar System, which is a collection of planets and satellites, comets, and other objects all of which orbit our parent star.

The Sun itself is but one of around 100,000 million stars that make up the huge system we call the Galaxy. We see the Galaxy as a pearly band of light crossing the sky. As we look along the plane of the Galaxy we see

imagination. A ray of light, travelling at around 300,000 km (186,000 miles) every second, would take a little over a second to cross the gulf between the Earth and Moon, and a little over five hours to reach the outermost-known planet Pluto. This same ray of light would take over four years to reach even the nearest star, a staggering 100,000 years to cross the Galaxy and over eight million years to reach M81. Yet the group of galaxies of which M81 is a member is regarded as a neighbour of the Local Group, the collection of galaxies of which our own is a member.

the combined light from untold thousands of stars as a luminescent band stretching completely around the sky. The spiral galaxy of which our Sun is a member resembles in many ways the spiral galaxy M81 (NGC (3031) in Ursa Major (see *above*). Spiral arms radiate out from a huge central bulge, our Solar System being located within one of these arms. It is from here that we gaze out upon the Universe.

The view we get is mind boggling. The vast distances between even our so called neighbours in space defy the

ABOVE The spiral galaxy M81 (NGC 3031) in Ursa Major is similar in many ways to our own Galaxy. This galaxy is regarded as a neighbour of the Local Group, a collection of galaxies of which our own is a member, yet a ray of light would take over eight million years to travel the distance between the Earth and M81.

Astronomers are constantly gazing into the depths of the Universe through huge telescopes, these being situated at remote, pollution-free locations across the globe. One such is the Kitt Peak National Observatory in Arizona. With such large telescopes available to professional astronomers, you may think that the amateur has little part to play. Yet nothing could be further from the truth. While it is true that many avenues of exploration are solely the role of the professional, there is much the amateur can do and see.

The beautiful cloud belts of Jupiter, the glorious rings of Saturn and barren landscape of the Moon are but some of the wonders our own Solar System has to offer. Beyond this we see multiple stars, star clusters and nebulae, while beyond the confines of the Galaxy are other galactic systems. Examples of each of these objects are within the light grasp of small telescopes, binoculars or even the naked eye. Descriptions of many are given in this book including details of how to track them down

ABOVE The Kitt Peak National Observatory in Arizona is seen here backlit by the setting Sun. One of the eyes used by the astronomers here is the 4-m (13-ft) diameter Mayall Telescope, housed in the dome seen a little way to the right of centre of this picture. It was this telescope that captured the accompanying image of M81.

and what you can expect to see when you turn your gaze towards them.

The Universe is indeed huge, the distances unimaginable, and the sizes of some of the objects we see beyond our comprehension. Yet this adds to the wonder of astronomy. It is a sobering thought that a ray of light may take many tens of thousands of years to cross the seemingly-tiny patch of light you see in your telescope, and that the light from the object you are looking at may have taken tens or even hundreds of millions of years to complete its journey to your inquisitive eye.

BINOCULARS

Binoculars are, in effect, a pair of small refracting telescopes joined together. Observation is carried out using both eyes simultaneously, which is far less tiring than keeping one eye closed for prolonged periods when using a conventional telescope. Binoculars are also relatively inexpensive, versatile, easy to use and highly portable. This is why they are the first choice for many amateur astronomers.

CHOOSING BINOCULARS

Binoculars are classified according to their magnification and aperture, a pair of 10 × 50 binoculars giving 10× magnification with 50-mm diameter objectives. As with telescopes, the larger the aperture, the more light will be gathered and the brighter will be the image. The combination of aperture and magnification should be considered when buying binoculars. Instruments with large objectives are heavy, and difficult to hold steady for more than a few moments at a time, as are those with magnifications of 12× or more. Hand-held binoculars should really have no more than 7× magnification, although this is up to individual choice. Binoculars with larger magnifications give correspondingly narrower fields of view and are consequently harder to hold steady during use.

As discussed elsewhere (see Observing the Sky, pages 12-13), the maximum diameter of the fully dark adapted pupil is around 7 mm (¼ in). Ideally, the exit pupil of the binoculars should match this so that all the light emerging from the binocular eyepieces is utilized by the eye. The exit pupils are the images formed by the binoculars. They can be seen by holding the binoculars away from your eyes with the objectives directed towards a light source and will be visible as a disc of light in each eyepiece. The diameter of the exit pupil is equal to the aperture divided by the magnification. Thus a pair of 12 × 50 binoculars will have exit pupils of 50 mm ÷ 12 — just over 4 mm (⅛ in). The optimum exit pupil, as we have seen, is around 7 mm (¼ in). This is the size given by 7 × 50 binoculars, which is one reason for this being an ideal first choice of size. 7 × 50 binoculars also have good light-gathering power, and are quite light and easy to hold steady during use.

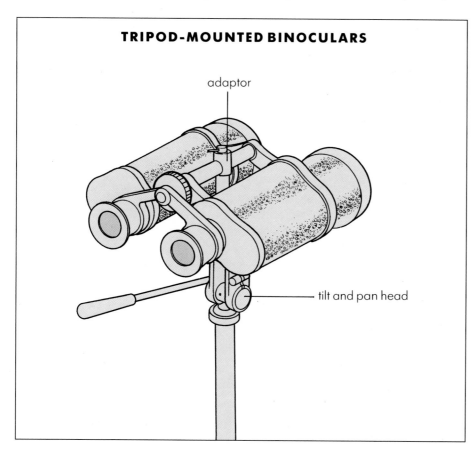

TRIPOD-MOUNTED BINOCULARS

adaptor

tilt and pan head

LEFT Mounting binoculars on a tripod, using a special adaptor, ensures that they are held firmly and rigidly in place; moreover, both hands are left free. Ideally, the tripod should have a tilt and pan head so that the sky can be viewed at or near zenith. The tripod should be between 1.5 m (5 ft) and 1.8 m (6 ft) tall and sturdy. Although mounting any type of binoculars will aid viewing, it is most necessary for binoculars with magnifications of 12× or more; the narrow fields of view of high magnitude instruments make them difficult to hold steady for prolonged periods.

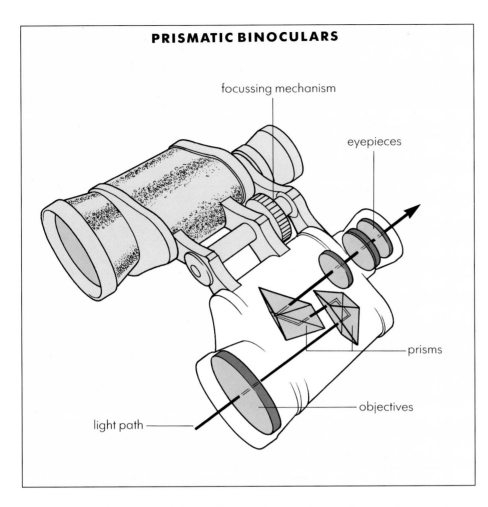

PRISMATIC BINOCULARS

focussing mechanism

eyepieces

prisms

objectives

light path

LEFT This cut-away view of prismatic binoculars shows the numerous optical surfaces encountered by the incoming light. The folded light path results in light and compact binoculars.

One main advantage with binoculars is that they can be examined thoroughly before you part with your money. Before you buy, make sure that they come with a money-back guarantee in the event that they prove unsatisfactory. Initial checks should include whether they are well balanced and comfortable to hold. The focussing mechanism should run smoothly and firmly, and it should remain in place after adjustment. Scrutinize the objectives and eyepieces to make sure they contain no scratches or chips and examine the binoculars in general to confirm the overall standard of workmanship. Any dents in the binocular barrels may indicate previous rough handling and may hint at further, internal damage. Shake the binoculars to see whether they rattle. The main barrels should be rigid with no play, as should the eyepiece barrels when fully extended.

A look through the binoculars should present a field of view that is sharp at the edge. The image you see should also be sharp and clear almost to the edge. There are few binoculars which give a completely sharp image all across the field of view. A star should focus to a distinct point. Any binoculars that do not deliver such an image should be discarded as should those that display excessive amounts of false colour.

The exit pupils should be round; greyish 'squared-off' edges on the discs indicate that not all the light entering the objectives is getting past the prisms to the eyepieces. The actual light transmission of the optical components is also an important consideration. Most binoculars available today have anti-reflection coatings on the outside of the objectives, thus improving light transmission and increasing contrast. Standing with your back to a light source and looking at the reflection of that light source in the binocular objectives will reveal whether the lenses are indeed coated. A coloured reflection indicates the presence of a coating, white reflec-

tions signifying uncoated optics. A second reflection should be visible from the inside surface of the lenses. Is this surface coated? Similar tests can be made at the eyepiece end. Discard binoculars that have no optical coatings.

TIP

AFTER CARE

Look after your binoculars and they will look after you. Avoid touching the optical surfaces with your hands. Clean only the outside surfaces of the objectives and eyepieces, either by blowing with a syringe or by gently brushing with a camel hair brush. The strokes should be light; shake the brush after each stroke to dislodge any dust particles that may have been gathered.

Make sure you get a full set of protective caps for the binoculars, together with a sturdy carrying case. You should also minimize the risk of dropping the binoculars by acquiring the habit of wearing the carrying strap whenever you are observing.

TELESCOPES

A telescope is an expensive item to invest in, so consider the following before buying. First, the beginner should be well-acquainted with the night sky and have gained experience of naked-eye and binocular observation before advancing to telescopes. When ready, carefully consider what type of instrument is required. Most first-time buyers need an all-purpose instrument. Others may wish to observe particular types of deep-sky object – double stars or star clusters, perhaps – which will limit the choice of suitable telescopes.

TYPES OF TELESCOPE

There are three main types of telescope – refractors, reflectors and catadioptrics – each employing a different optical system.

REFRACTORS use a lens, or objective, to collect light; the objective is designed so that the light is refracted, or bent, to a focus.

Light entering a CATADIOPTRIC system first passes through a corrector lens before being brought to a focus via primary and secondary mirrors. As can be seen in the diagram, the light is diverted through a hole in the main telescope mirror. This produces a folded light-path, which makes the catadioptric telescope a very compact and highly portable instrument. REFLECTORS also have a folded light-path although the end result is nowhere near as compact as that of the catadioptric system. Reflectors use a mirror to collect light; the special concave shape reflects the light back up the tube to a small secondary mirror, or flat, mounted along the central axis of the telescope tube which reflects the light to the focal point. This secondary mirror does cut out some of the incoming light, although not enough to cause a serious loss in image brightness.

NOTE: The distance between the lens or mirror and the focal point is the focal length of the telescope.

LENSES

In all types of telescope a tiny image of the object being observed is formed at the focal point. This image is magnified by an eyepiece. This is a lens, or compound set of lenses, mounted in a small tube designed to fit into the end of the telescope. Generally speaking, telescopes are equipped with a number of eyepieces to give a range of magnifications. The magnification given by an eyepiece is calculated by dividing its focal length into the focal length of the main telescope. For example, using a telescope of focal length 600 mm with an eyepiece of focal length 25 mm will give a magnification of $600 \div 25$ or $24\times$. There are several different types of eyepiece, each of which is designed to suit particular types of telescope. One of the most popular types of eyepiece is the orthoscopic, which gives a wide field of view and delivers fine images, ideal for rich-field telescopes.

APERTURE VERSUS MAGNIFICATION

The diameter of a telescope lens or mirror governs that telescope's light-gathering power. The larger the aperture, the more light it gathers, and therefore the brighter the resulting image, and the fainter the objects that can be successfully observed through it. When purchasing a telescope, the size of aperture is the main consideration.

The usefulness of a telescope is governed by its 'maximum useful magnification' (otherwise known as the 'optimum magnification'). Some manufacturers boast very high magnifications for their telescopes, although in the vast majority of cases these figures are meaningless.

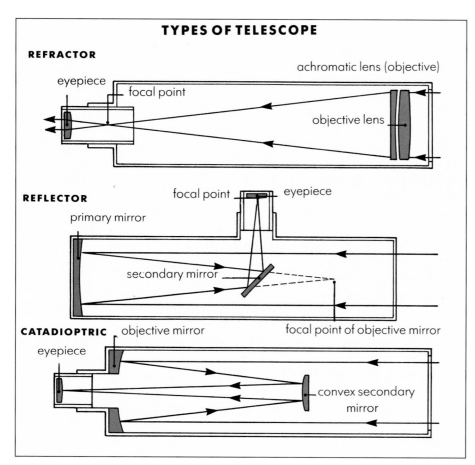

TYPES OF TELESCOPE

REFRACTOR
- eyepiece
- focal point
- achromatic lens (objective)
- objective lens

REFLECTOR
- primary mirror
- focal point
- eyepiece
- secondary mirror

CATADIOPTRIC
- objective mirror
- focal point of objective mirror
- eyepiece
- convex secondary mirror

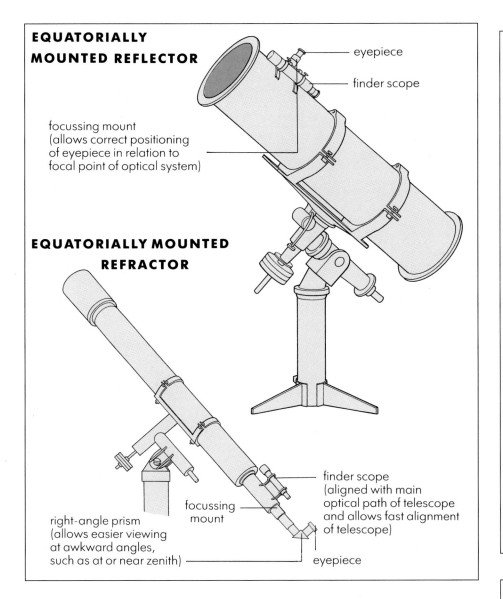

EQUATORIALLY MOUNTED REFLECTOR

eyepiece

finder scope

focussing mount
(allows correct positioning
of eyepiece in relation to
focal point of optical system)

EQUATORIALLY MOUNTED REFRACTOR

right-angle prism
(allows easier viewing
at awkward angles,
such as at or near zenith)

focussing
mount

finder scope
(aligned with main
optical path of telescope
and allows fast alignment
of telescope)

eyepiece

REFRACTORS AND FALSE COLOUR

Not all wavelengths of light are refracted equally. As Isaac Newton discovered during the 17th century, white light is a blend of all colours of the spectrum ranging from long wavelength red through to short wavelength blue. These are brought to focus at different points. Long wavelength red is refracted less than short wavelength blue, red light being brought to a focus at a greater distance from the objective (see diagram *below*). This effect, which does not occur in reflectors, due to the fact that mirrors reflect all wavelengths equally, is known as 'chromatic aberration' and produces false colour around the image of a star or other bright object.

Although chromatic aberration can never be completely eliminated, the problems are overcome to a great extent by constructing the telescope objective from several components. Each of these is made from a different type of glass, the overall effect being to cancel out much of the chromatic aberration thereby reducing the amount of false colour present in the image. The fact that objectives contain several components, each of which has two faces to be figured, tends to make refractors more expensive, aperture for aperture, than reflectors. The mirrors used in refractors require only one surface to be worked.

The highest magnification that can be successfully used depends upon the amount of light available and therefore upon the aperture of the telescope. The higher the magnification used, the fainter the image delivered by the telescope. This is because, as magnification increases, the area being observed gets progressively smaller, the less light emerges from it, and consequently, the dimmer is the resulting image. The principle that the greater the telescope aperture, the more light it can gather, is the basis for dark adaptation of the eyes (see Observing the Sky, pages 12-13). Larger apertures mean accessibility to fainter objects — high magnifications being practical only with telescopes of suitably large apertures.

Claims that a 60-mm (2.5-in) aperture of focal length 900 mm telescope will magnify 150× may be correct in theory. After all, using an eyepiece of 6 mm focal length will delivery that magnification. However, the resulting image will be so dim that little, if any, detail will be seen. 'Maximum useful magnification' is equal to around 50× per 25 mm (1 in) of aperture. Therefore, the maximum useful magnification for the 60-mm (2.5-in) telescope would be 125×. To be able to effectively use a magnification of 150× you would need a telescope of at least 75-mm (3-in) aperture. The optimum magnification should legally be given with the description of the telescope. If it is not, and the seller is unable to provide the figure, walk out of the shop!

CHROMATIC ABERRATION PRODUCING FALSE COLOUR

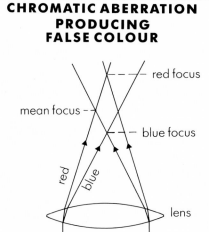

red focus

mean focus

blue focus

red

blue

lens

light rays

OBSERVING THE SKY

There are many different objects that are accessible to the amateur astronomer. Naked-eye observers can, among other things, identify the constellations and the brightest stars, and watch the motions of the planets through the sky. Binoculars extend the range of sight enormously, bringing dozens of star clusters, nebulae and other deep-sky objects into view, while telescopes reveal much more.

Knowing where to look and what you will see when you find what you are looking for enhances the rewards astronomy can bring. The important thing to remember is that you will not see the visually stunning sights presented by professional telescopes. The clusters, nebulae and galaxies portrayed in astronomy books often bear little resemblance to the view obtained by the amateur astronomer. Yet, although the objects within your visual reach usually appear as diffuse patches of light, the more time you spend observing the sky, the more detail you will be able to make out. A galaxy, which at first appears as nothing more than a patch of light may, as you gain experience, show traces of dust lanes against the galactic disc or perhaps hints of spiral arms radiating from a central bulge. The reward for persistence is that you will actually *learn* to see!

This book contains whole-sky charts together with wide-field charts of particular constellations and more detailed finder charts, which will help you track down the objects you seek. These finder charts make use of the star-hopping method of deep-sky object location (see How To Use The Star Charts, pages 66-67). You will find some objects easier to spot than others. If you have difficulty finding an object, try another night.

DARK-ADAPTATION

There are a number of points to remember when observing the night sky. First and foremost is that your eyes should be fully dark-adapted. When you are in a brightly lit room, the pupils of your eyes are nearly closed. Under dark conditions they expand to allow more light into the eye. It is rather like adjusting the diaphragm in a camera to suit different levels of light. The wider the pupils, the more light reaches the retina. This adjustment, which amounts to a normal maximum of around 6 or 7 mm (¼ in), is called dark-adaptation, your eyes adapting to the lower levels of light available under dark skies. Full dark-adaptation can take up to an hour, although the eye becomes fairly effective after ten minutes or so.

EQUIPMENT AND DRESS

Once your eyes have become dark-adapted, do not ruin matters by using a bright light to illuminate your star charts! Any illumination used at the telescope should be made by a red light, the wavelengths of which produce least reaction in the eyes.

RECORDING EQUIPMENT

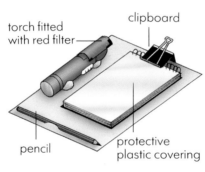

torch fitted with red filter

clipboard

pencil

protective plastic covering

Observing notes should be on paper mounted on a clipboard with a plastic sheet to cover the paper when not in use, preventing damp settling on it. Pencils are a definite asset, pens often ceasing to function under damp conditions!

Dress in an appropriate manner. It is surprising how cold it can get when you are outside for an extended time, even on summer nights. Wear a thick coat (with plenty of pockets for such things as pencils, spare eyepieces and so on), warm trousers, scarf, and a pair of gloves; the fingerless type of glove is useful as they allow for easier handling of eyepieces, with less chance of dropping them. Warm undergarments should also be worn. Finally, it should be pointed out that most body heat is lost through the head and feet. Some sort of hat or hood should be worn as well as warm socks, and comfortable shoes or boots. Tight-fitting footwear will make things worse by restricting circulation in the feet.

THE CELESTIAL SPHERE

To all intents and purposes, the stars and other objects we see are set against the imaginary sphere of the sky. This sphere is called the celestial sphere and it contains a number of reference points. First of these are the celestial poles, which are the points on the celestial sphere which lie directly above the north and south terrestrial poles (see North Circumpolar Stars, pages 68-69, and South Circumpolar Stars, pages 118-119). The celestial equator is a projection of the Earth's equator onto the celestial sphere. Extending fully around the sky, it lies equidistant from each celestial pole.

Distances on the celestial sphere are measured in angles, the units of which are degrees, minutes and seconds of arc (see Double and Multiple Stars, pages 74-75). Although the stars are all moving through space, they lie at such immense distances that their motions are difficult to detect even over periods of many years. They are deemed to occupy fixed pos-

THE CELESTIAL SPHERE

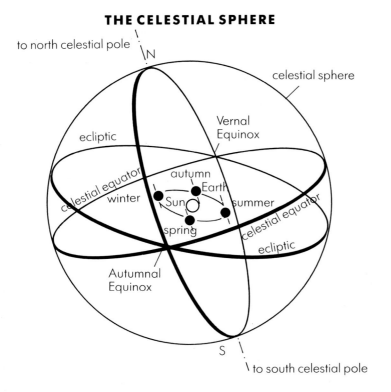

to north celestial pole

N

celestial sphere

ecliptic

Vernal Equinox

celestial equator

autumn

Earth

winter

Sun

summer

spring

celestial equator

ecliptic

Autumnal Equinox

S

to south celestial pole

itions on the celestial sphere. However, the Moon, planets, comets, and other members of the Solar System move through the sky as they orbit the Sun. Their positions against the backdrop of stars are constantly changing as a result. The Sun also appears to move although, unlike the motions of the planets, that of the Sun is a result of our planet orbiting it. As the Earth travels around our star, we see the Sun from a slightly different position each day. The effect of this is that the Sun appears to shift against the backdrop of stars. The Earth makes one complete orbit of the Sun in a year, the Sun appearing to travel completely around the sky during this time. The imaginary path taken by the Sun is known as the ecliptic, and the two points at which it crosses the celestial equator are the Vernal and Autumnal Equinoxes.

The accompanying diagram (*above*) shows the main parts of the celestial equator and also the position of the Earth at different times of the year. It is clear from the diagram why we see different stars at different times of the year. During northern summer, for example, the daytime sky is the region to the left of the diagram, this being the area of sky con-

taining the Sun. The night-time sky is that occupying the right side of the diagram. Six months later, the situation is reversed, the Sun occupying what was the night-time sky during winter, leaving the stars on the left of the diagram in a dark sky.

It can also be seen that the position of the Sun relative to the celestial equator varies throughout the year, being located north of (above) the celestial equator in northern summer and south of (below) the celestial equator in northern winter. These variations are the main cause of the seasons on Earth; the higher the Sun is in the sky, the greater will be its warming effect. Note also that while it is summer in the northern hemisphere, those below the equator experience winter, similar reversals taking place for the other seasons. As a result, northern-hemisphere observers see basically the same section of the celestial sphere at night during northern winter as those in the southern hemisphere see during southern summer. For example, Orion (see Deep Sky Objects: Northern Winter/Southern Summer 2, pages 80-81) is classed as a winter object for northern-hemisphere observers and a summer object for those in the southern hemisphere.

FACT

TAKING IT FURTHER

This book is intended to be a guide for those who are not content merely to look at the sky, but who want to make practical observations. Several organizations offer guidance for amateur astronomers who want to carry out useful work; anyone who wants help in a specific area of interest is advised to contact one of the following organizations. A number of these organizations encourage observation within particular areas of interest, such as meteors, solar, planetary and so on. For further details write, not forgetting to enclose a SAE or reply-paid coupons:

BRITISH ASTRONOMICAL ASSOCIATION
Burlington House, Piccadilly, London, W1V 9AG, England

JUNIOR ASTRONOMICAL SOCIETY
36 Fairway, Keyworth, Nottingham, NG12 5DU, England

FEDERATION OF ASTRONOMICAL SOCIETIES
c/o Christine Sheldon, Whitehaven, Maytree Road, Lower Moor, Pershore, Worcestershire, WR10 2NY, England

AMERICAN ASSOCIATION OF VARIABLE STAR OBSERVERS
25 Birch Street, Cambridge, MA 02138, USA

AMERICAN ASTRONOMICAL SOCIETY
2000 Florida Avenue NW, Suite 300, Washington, DC 20009, USA

ASSOCIATION OF LUNAR AND PLANETARY OBSERVERS
8930 Raven Drive, Waco, TX 76712, USA

ASTRONOMICAL SOCIETY OF THE PACIFIC
390 Ashton Avenue, San Francisco, CA 94112, USA

ROYAL ASTRONOMICAL SOCIETY OF CANADA
McLaughlin Planetarium, 100 Queens Park, Toronto, Ontario, Canada M5S 2C6

BRITISH ASTRONOMICAL ASSOCIATION
New South Wales Branch, PO Box 103, Harbord, New South Wales 2096, Australia

ROYAL ASTRONOMICAL SOCIETY OF NEW ZEALAND
PO Box 3181, Wellington, New Zealand

LIGHT POLLUTION

O f all the different types of pollution currently threatening our planet, perhaps the least publicized (outside astronomy circles) is light pollution. What is light pollution? The answer to this is obvious to those who, in an attempt to 'get away to darker skies', have taken their telescopes out into the countryside in order to benefit from the superior observing conditions. A glance towards a populated horizon from such locations will reveal the all-too-familiar sky glow hanging over the nearby towns and cities.

· · · ● ● ● ● ●

THE EFFECT OF LIGHT POLLUTION

Sky glow is produced by waste-light from streetlights and other forms of outdoor illumination which escapes upwards where it is scattered by dust particles in the atmosphere. The resulting glow makes it difficult or impossible to observe the sky in that direction. Because of light pollution, the pearly glow of the Zodiacal Light, Gegenschein and even the Milky Way, are swamped in urban skies. The wide-angle photograph (see *below*) of the Milky Way, taken from Mount Graham, Arizona, clearly illustrates the effect of sky glow from populated areas. The glare around the edge of the picture comes from Tucson and Phoenix and rivals the Milky Way for prominence. To catch a glimpse of these elusive objects, city-based astronomers need to trek out to rural areas where the darker skies will permit their observation.

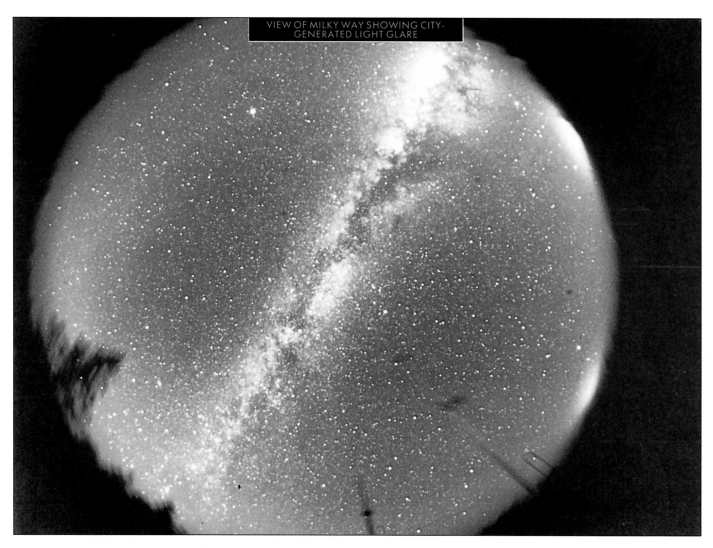

VIEW OF MILKY WAY SHOWING CITY-GENERATED LIGHT GLARE

THE GROWTH OF THE PROBLEM

The dramatic growth in urban light output is clearly indicated in the pair of photographs (see *right*) recording the light output of Tucson, Arizona, between 1959 (top) and 1980. Sadly, even professional observatories have suffered because of light pollution. At Mount Wilson in California, the 2.5-m (100-in) Hooker telescope had to be taken out of service, the reason being that light pollution from nearby Los Angeles largely ruined the telescope's observation of faint stars and galaxies. Modern technology and the spread of human population has created a light-polluted environment around the once superbly dark site.

THE ALTERNATIVES

Both amateur and professional astronomers have for a number of years been trying to halt the steadily growing effects on our skies of outdoor lighting. Not only is this detrimental to astronomers, but also to the tax-payer, whose money is wasted through this grossly inefficient way of using valuable energy resources. After all, these lights are intended to illuminate the ground, not the sky. Indeed, economics is the astronomer's major weapon in the fight against light pollution. Better designs for streetlighting would not only give astronomers an easier time, but save the public fuel-consumers considerable amounts of money. Astronomers have had an uphill struggle so far, although there have been a few successes. For instance, the observatories on Kitt Peak and Mount Hopkins, both in Arizona, and Palomar Observatory in California, have been saved as a result of the enforcement of lighting regulations in nearby urban areas. Installation of certain types of lighting is also beneficial to astronomers. The photograph shown here (*right*) is of a parking lot that has been equipped with low-pressure sodium lighting which emits at just a single yellow frequency, thereby ensuring that the rest of the visible light region is 'dark' for astronomers.

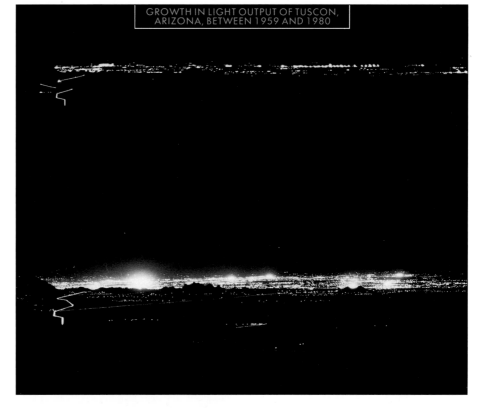

GROWTH IN LIGHT OUTPUT OF TUSCON, ARIZONA, BETWEEN 1959 AND 1980

LOW-PRESSURE SODIUM LIGHTING

With a diameter of 1,392,530 km (865,000 miles) and a mass totalling 98 per cent of the total mass of the Solar System, the Sun governs our region of space. This mighty cosmic power house pro-duces practically all the light and heat essential for life on our planet. Like most other stars, the Sun is composed mainly of hydrogen and helium, although its brilliance is merely a result of its proximity to us.

THE COMPOSITION OF THE SUN

The Sun's proximity enables astronomers to study it in far more detail than any other star, and by continually monitoring solar activity we are able to improve our knowledge of stellar behaviour and how stars in general produce their energy. The energy emitted by the Sun (and other stars) emerges from the outer visible sur-face, or 'photosphere'. The temperature of the photosphere is around 5,500°C (9,900°F), a value which, although high by terrestrial standards, is little in com-parison to the temperature of 15 million °C (27 million °F) at the Sun's core. It is here, deep inside the Sun, that all the solar energy is produced. This colossal temperature, coupled with the pressure of the overlying gases in excess of a third of a million times the atmospheric pres-sure at the Earth's surface, perpetuates a continuous series of nuclear reactions during which four hydrogen nuclei are fused together to form one nucleus of helium. This process of hydrogen burn-ing, known also as the proton-proton reaction, results in the conversion to helium of a colossal 700 million tonnes of hydrogen every second! During the process, around 0·7 per cent of the mass involved in the reactions is left over. At first this may seem insignificant, although calculation shows that it is equal to al-most 5 million tonnes. This left-over mass slowly makes its way to the surface of the Sun by both radiative and convective processes. Initial movement is through radiation whereby the photons of energy are passed between individual atoms and electrons. These motions, although random, act in an overall direction out-wards from the core.

Beyond the radiative zone convection takes over, during which process hot gas is bodily conveyed to the surface. As the hot gas circulates, reaching the surface at speeds of up to 1,800 km (1,120 miles) per hour, its energy es-capes as light and heat. The cooler gas then descends once more to the solar interior. This emergence, cooling and sinking of gas can be observed by closely studying the photosphere which, as examination will show, has a granular appearance. These granulations are actually turbulent cells with diameters of around 1,000 km (620 miles). They com-prise bright central regions, the areas within which hot gas is emerging, sur-rounded by dark lanes or boundaries, showing where the gas, cooled through several hundred degrees, is falling back into the Sun.

After emerging from the photosphere, all solar energy passes through the chromosphere on its way out through the corona, the latter being loosely labelled as the Sun's atmosphere. Neither the chromosphere or corona are bright enough to be seen against the over-whelming glow from the photosphere and are only visible through specialized equipment or during a total solar eclipse when the bright solar disc is temporarily blotted out by the Moon. The corona stretches millions of kilometres (miles) into space and, like the chromosphere, is heated from below by as-yet unex-plained processes. The rise to the high temperatures of the chromosphere (in excess of 8,000°C/14,500°F) and the corona (over 1,000,000°C/1,800,000°F) are produced in the transition zone, a narrow region located between the chromosphere and the base of the corona.

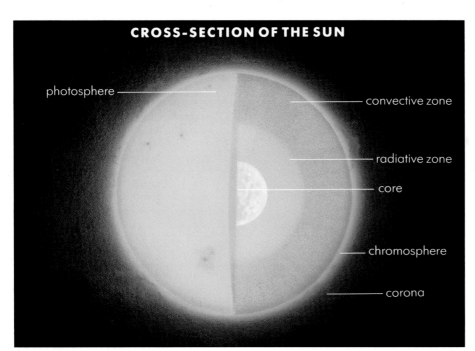

CROSS-SECTION OF THE SUN

photosphere

convective zone

radiative zone

core

chromosphere

corona

COMPARABLE STARS

The Sun is close by astronomical standards, the nearest star to us (other than the Sun) being the dim red dwarf Proxima Centauri which shines from over a quarter of a million times the distance. Proxima Centauri is a member of the Alpha Centauri star system, Alpha Centauri being the left star of the bright pair of stars seen near the bottom margin of this photograph (see *right*). Lying at a distance of 4·34 light years, Alpha Centauri is the nearest of the bright stars and is similar in many ways to the Sun. Were the Sun to be placed at the same distance as Alpha Centauri it would shine as a first magnitude star and have an apparent brightness roughly half a magnitude fainter than Alpha Centauri itself.

CONSTELLATIONS OF CENTAURUS AND CRUX

SUNSPOTS

Appearing as dark patches on the surrounding bright photosphere are sunspots, such as this unusual spiral-shaped sunspot (see *right*) photographed on 19 February, 1982. Sunspots are the best-known type of solar feature and many are clearly visible through amateur telescopes. They take the form of depressions in the photosphere and only appear dark due to being cooler than the surrounding photosphere. Typical sunspots have two distinct regions: a dark, central umbra, with a temperature in the region of 3,800°C (6,900°F), surrounded by a lighter, warmer penumbra. There is no strict rule governing the sizes of sunspots, their diameters being anything up to several tens of thousands of kilometres (miles) or more. Some spots are so large as to become observable with the naked eye. Sunspot activity takes place over a well-defined cycle with periods of maximum sunspot activity occurring every 11 years. Often, highly luminous clouds called faculae can be seen near sunspots. Comprised primarily of hydrogen, the frequency of their appearance is governed by the 11-year solar cycle. They appear above the regions where sunspots are due to appear and often remain after the associated sunspot has disappeared.

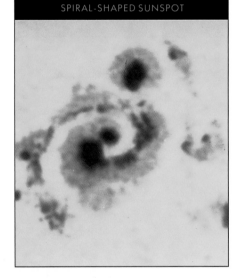

SPIRAL-SHAPED SUNSPOT

SOLAR FLARES

A solar flare (*right*) is a type of feature which erupts from sunspot regions and occurs both in the chromosphere and in the lower corona. Having lifetimes of anything between a few minutes and several hours, these bright filaments of hot gas can attain temperatures of several million degrees over short periods. The appearance of large flares can produce interference in radio communication, although the resulting ejection of the increased amounts of energized particles leaving the Sun can result in increased auroral displays.

SOLAR FLARE

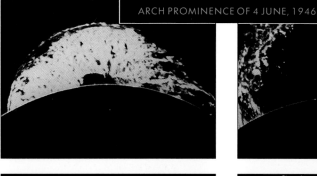

ARCH PROMINENCE OF 4 JUNE, 1946

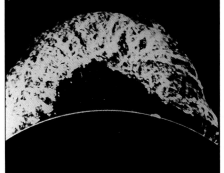

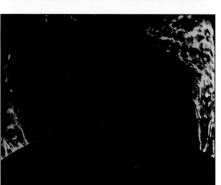

PROMINENCES

Huge columns of hot gas can sometimes be seen erupting from the solar surface. These are prominences and there are two distinct types: quiescent prominences, which are fairly docile and, once formed, tend to remain suspended over the photosphere for days on end; and eruptive, or active prominences, such as this one recorded on 4 June, 1946, which are much more spectacular. This sequence of photographs (*left*) shows the growth of a huge arch prominence over a period of one hour. As with other eruptive prominences, this one was formed from gas leaving the solar surface at speeds of up to 1,000 km (620 miles) per second. As can be seen, their forms can alter from minute to minute and can attain heights of over 500,000 km (300,000 miles) above the solar surface.

THE ULYSSES PROBE

The joint ESA/NASA Ulysses probe was launched from *Discovery* during Space Shuttle mission STS-41 in October 1990. Its aim is to study the Sun's polar regions and also to examine the interplanetary environment in the as-yet unexplored regions above the solar poles. Up until now, all observation of the Sun has been carried out from within, or close to, the plane of the Earth's orbit around the Sun, from where we are unable to effectively study those regions of the Sun at or near its poles. Ulysses, however, will be providing new scientific opportunities.

Following its launch, Ulysses headed for an encounter with Jupiter in February 1992, as can be seen from the accompanying flight profile (*inset right*). During its fly-by of the planet (see artist's impression *right*) it will be at its maximum distance of 800 million km (500 million miles) from the Sun. Ulysses will utilize the powerful Jovian gravitational field to swing it out of the main plane of the Solar System and southwards towards a passage over the solar south polar region between June and October 1994, when it will be within 320 million km (200 million

miles) of the solar surface. It will then head northwards, passing over the solar equator in February 1995, making its north polar pass between June and September 1995. Ulysses is not fitted with optical cameras; its equipment being

designed rather to examine solar radiation, including X-rays and radio emissions. The structure and extent of the solar magnetic field will also be explored – from much higher solar latitudes than ever before possible.

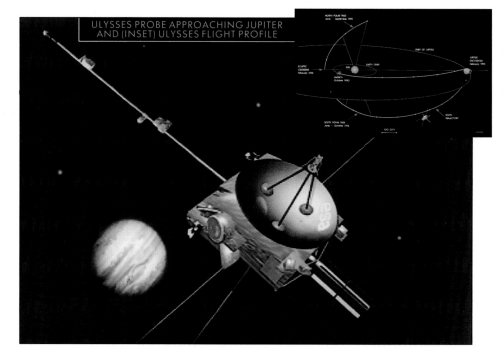

ULYSSES PROBE APPROACHING JUPITER AND (INSET) ULYSSES FLIGHT PROFILE

The Sun is the most visually accessible celestial object available to astronomers, yet its very accessibility makes it a danger to the unwary backyard observer. Gazing at the Sun without taking suitable precautions, even for a short instant, can lead to damaged eyesight. There is actually little that the naked-eye observer can see other than the occasional large sunspot. However, only search for these when the Sun is very low over the horizon and seen through a layer of haze. Even then, a dark filter **must** be used.

When the Sun is seen near the horizon, it often looks markedly flattened. This effect is caused by bending, or refraction, of the Sun's light through the Earth's atmosphere, the effects of refraction being greater the nearer the Sun (or other celestial object) is to the horizon. The Sun appears flattened simply because the lower limb of the solar disc is refracted more than the upper, the Sun appearing 'squashed' as a result!

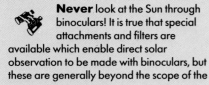 **Never** look at the Sun through binoculars! It is true that special attachments and filters are available which enable direct solar observation to be made with binoculars, but these are generally beyond the scope of the backyard astronomer. Binoculars can be used to project an image of the Sun onto a screen (in a similar way to solar projection carried out with a telescope and described below), but this is not recommended. The lenses of most binocular eyepieces are secured together with a special cement which may well melt under the concentrated heat from the Sun. The same is true of most binocular object glasses which, unlike the air-spaced components of telescope objectives, are cemented together.

Telescopes offer the best opportunities for the would-be solar observer, although direct observation can be dangerous and must **never** be carried out. The recommended method for observing the Sun is to *project* an image of the solar disc onto a screen supported about 30 cm (12 in) from the eyepiece. The telescope is aligned with the Sun and used to focus the image. Telescopes often come equipped with a solar-projection screen, but it is a straightforward job to construct your own if required. A sun shade should also be fastened to the telescope, as shown. This will cut out any direct sunlight from the screen, making the image brighter and easier to work with.

To align the telescope, set the screen up as shown and turn the instrument in the general direction of the Sun. Then slowly move the tube around until you see the image on the screen. Once you reach this stage, the image can be brought roughly into focus by adjusting the distance of the screen from the eyepiece. Fine focussing is then carried out by racking the eyepiece. Under **no circumstances** use the telescope finder to align the telescope! If your finder has a lens cap, fit it prior to commencing observation.

A good image size is 150 mm (6 in) in diameter and a useful idea is to make a number of disc-blanks comprised of pieces of card with circles of this diameter drawn on them. These cards are then attached to the screen each time an observation is made and the projected disc centred on it. The card can also contain details such as name of the observer, date, time, observing conditions and instrument size.

Solar features that are visible when using solar projection include granulation on the solar disc, which may be seen when using magnifications of 100× or more. However, granulation is difficult to see and excellent observing conditions are normally required.

By far the most striking features are sunspots, which appear as dark patches on the projected disc. If you see a sunspot, and repeat your observations over a period of days or weeks they will appear to cross the solar disc. This is an effect of the Sun's axial rotation, the spots being carried round the visible disc of the Sun as our star spins on its axis. The spots may change in appearance, growing or shrinking as they age. Drawing the spots will enable you to see their changing shapes. Also, marking their positions from day to day will enable you to follow them as they cross the solar disc.

Bright faculae may also be seen. Occurring above the photosphere, they are difficult to pick out when superimposed against the solar disc, and are most readily visible when near the limb. Faculae often precede sunspots and, if seen, a watch should be kept on the area in question as a sunspot or sunspot group may well make an appearance. When looking at the solar limb, you may notice that the photosphere appears darker in this region. This so-called 'limb darkening' is caused by the light from the photosphere having to pass obliquely through a greater depth of the solar atmosphere, its light being dimmed as a result.

SOLAR PROJECTION

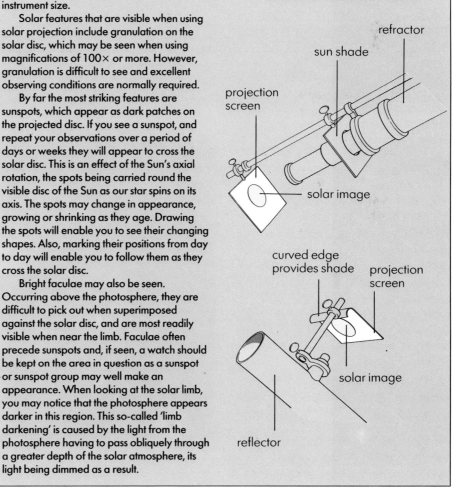

refractor
sun shade
projection screen
solar image

curved edge provides shade
projection screen
solar image
reflector

The Moon's axial rotation and orbital periods are identical, which means that the same lunar hemisphere is presented to the Earth at all times. However, the changing distance between the Earth and Moon results in the Moon's orbital speed varying slightly.

This in turn means that the rotation and orbital periods get slightly out of step with each other, producing a sideways 'wobble' of the Moon. This wobble, known as libration, permits us to view up to 59 per cent of the lunar surface over a period of time.

MOON DATA

EQUATORIAL DIAMETER (km/miles)	3,476/2,172
MASS (EARTH = 1)	0·0123
VOLUME (EARTH = 1)	0·0203
AXIAL ROTATION PERIOD (days)	27·32
ORBITAL PERIOD (days)	27·32
AVERAGE DISTANCE FROM EARTH (km/miles)	384,400/238,866
MINIMUM DISTANCE FROM EARTH (km/miles)	356,410/222,756
MAXIMUM DISTANCE FROM EARTH (km/miles)	406,697/254,186
MEAN DENSITY (g/cm³)	3·342

EARTH DATA

EQUATORIAL DIAMETER (km/miles)	12,756/7,927
AXIAL ROTATION PERIOD	23h 56m 04s
AXIAL TILT	23° 27'
ORBITAL PERIOD (days)	365·256
AVERAGE DISTANCE FROM SUN (km/miles)	149,600,000/ 92,961,440
INCLINATION OF ORBIT TO ECLIPTIC (°)	0·00
MEAN DENSITY (g/cm³)	5·517
NUMBER OF SATELLITES	1

THE LUNAR SURFACE

The lunar surface plays host to two main types of terrain, the bright, heavily cratered highlands standing out against the much darker, smoother lunar plains (see *right*). Prior to the invention of the telescope, astronomers believed that the dark areas were expanses of water. We now know differently, although the romantic names given to these areas, such as Mare Serenitatis (Sea of Serenity), Sinus Iridum (Bay of Rainbows) and Oceanus Procellarum (Ocean of Storms) are still used today.

The lunar plains were formed between 3 and 4 billion years ago as lava welled up from the lunar interior and filled the low-lying areas of the lunar surface. By this time, the main period of meteoritic bombardment that had peppered the surfaces of the Moon and other rocky planets had subsided. It is clear that the lunar maria were formed after the main period of meteoritic bombardment had subsided, these regions containing relatively few craters. This is in total contrast to the chaotic highlands which play host to various other features including mountains and valleys.

LUNAR PLAINS APPEAR DARK AGAINST HIGHLAND TERRAIN

There have been several theories as to the origin of the Moon. Following examination of the chemical composition of lunar rocks brought back by the Apollo astronauts, and taking into account similarities in density between the Moon and the Earth's outer layers, astronomers have come to the conclusion that the Moon formed as a result of the collision of an object some 6,500 km (4,000 miles) or so in diameter with the Earth during the early history of the Solar System. The collision resulted in the impacting object breaking up, its metallic core remaining with that of the Earth. The particles from its outer rocky mantle were pushed into orbit, eventually collecting together to form the Moon.

LUNAR PHASES

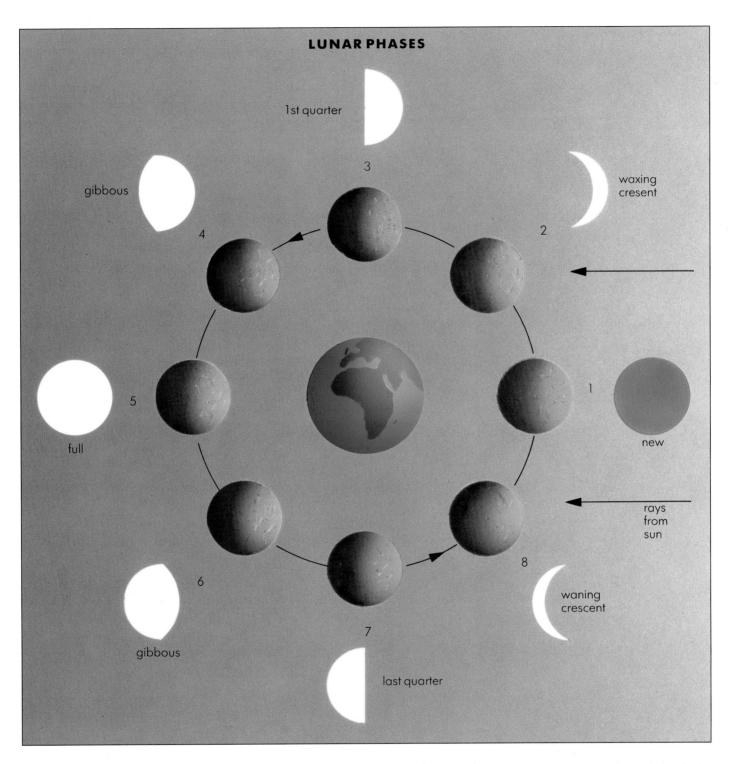

The phases of the Moon are caused by the constantly changing positions of the Sun and Moon relative to the Earth as shown on this diagram (*above*). The inner circle shows the changing position of the Moon, the outer circle illustrating how the Moon appears to us at each point in its orbit. New Moon occurs when the Moon is positioned between the Earth and Sun (1), the Moon's illuminated hemisphere facing away from the Earth. As the Moon moves along its orbit, more and more of its illuminated hemisphere becomes visible to us, the lunar phase gradually expanding through crescent (2), first quarter (3) and gibbous (4) to Full Moon (5). The sequence is then reversed until the following New Moon. It is important to note that, although the Moon takes 27·32 days to complete one orbit, the interval between successive New Moons is 29·53 days. This is because the Earth and Moon are travelling together around the Sun, producing a delay between the completion of one full lunar orbit and the realignment of the Earth, Moon and Sun.

THE MOON 2

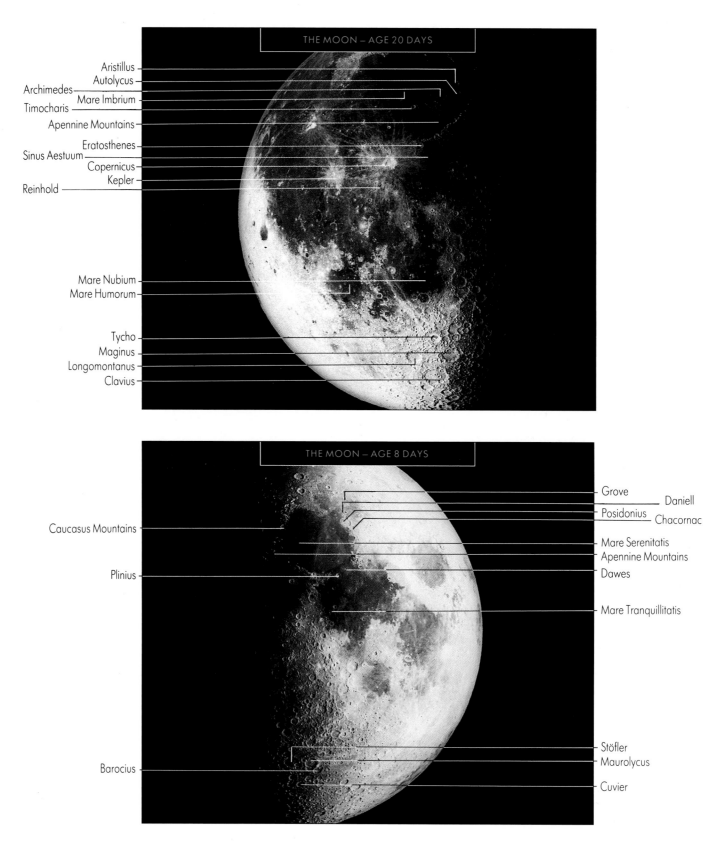

THE MOON – AGE 20 DAYS

Aristillus
Autolycus
Archimedes
Timocharis
Mare Imbrium
Apennine Mountains
Eratosthenes
Sinus Aestuum
Copernicus
Kepler
Reinhold

Mare Nubium
Mare Humorum

Tycho
Maginus
Longomontanus
Clavius

THE MOON – AGE 8 DAYS

Grove
Daniell
Posidonius
Chacornac
Caucasus Mountains
Mare Serenitatis
Apennine Mountains
Plinius
Dawes

Mare Tranquillitatis

Stöfler
Maurolycus
Barocius
Cuvier

REGION OF POSIDONIUS

This view (*above*) shows the area around the 100-km (62-mile) diameter crater Posidonius, seen here slightly above and to the right of centre. To its west (left) can be seen Mare Serenitatis (Sea of Serenity) while Mare Tranquillitatis (Sea of Tranquility) is visible across the bottom of the picture. Several small craters can be seen bordering Posidonius, including the crater Chacornac, the feature with disintegrated walls on the south-east wall of Posidonius. Chacornac is named after the French astronomer who discovered several minor planets.

To the north east of Posidonius are the pair of small but conspicuous craters, Daniell (closest of the pair to Posidonius) and Grove. The conspicuous pair of craters further to the north east are Hercules (left) and Atlas. The 67-km (42-mile) diameter Hercules is named after the legendary Greek hero, while Atlas, a 87-km (54-mile) diameter crater, is named after the mythological Titan of Greek legend.

The bright crater seen just to the east (right) of the terminator is Plinius. This 43-km (27-mile) diameter feature commemorates the Greek author Pliny who lived during the 1st century AD and who died during the destruction of Pompeii. Just to the east of Plinius is the smaller, 18-km (11-mile) diameter crater Dawes.

REGION AROUND CAUCASUS AND APENNINE MOUNTAINS

The Caucasus and Apennine mountain ranges straddle the borders between Mare Serenitatis, the circular mare in the lower half of picture (see *right*), and Mare Imbrium (Sea of Rains), spreading off the upper left margin of picture. The distinctive arc of the Apennine Mountains sweeps up from the lower left (southwest) into the 50-km (31-mile) wide flat channel between Mare Serenitatis and Mare Imbrium. The Caucasus Mountains are a direct continuation of the Apennines, seen here extending towards the upper (northern) borders of Mare Serenitatis.

The prominent 83-km (52-mile) flooded crater Archimedes, named after the Greek mathematician, is visible at left of the picture just above centre. To its right (east) are the bright pair of craters Aristillus (upper) and Autolycus. Aristillus measures some 55 km (34 miles) across while Autolycus is somewhat smaller at 39 km (24 miles) in diameter. The crater at right of the picture, on the eastern borders of Mare Serenitatis, is Posidonius (see page 23).

REGION OF CAUCASUS AND APENNINE MOUNTAINS

REGION OF TYCHO

REGION OF TYCHO

This photograph (*left*) shows the prominent crater Tycho, named after the Danish astronomer Tycho Brahe and seen below centre of the picture. Tycho has a central mountain towering 1·6 km (1 mile) above the crater floor while its terrace-like walls enclose an area some 85 km (53 miles) in diameter.

Tycho has a prominent system of rays. The ray systems of young craters such as Copernicus and Tycho are best seen at or near Full Moon. At these times they dominate the lunar surface and can easily be seen with the naked eye. The rays appear to have been formed from deposits thrown out during the formation of their associated craters, probably comprising material deposited on the lunar surface following impact.

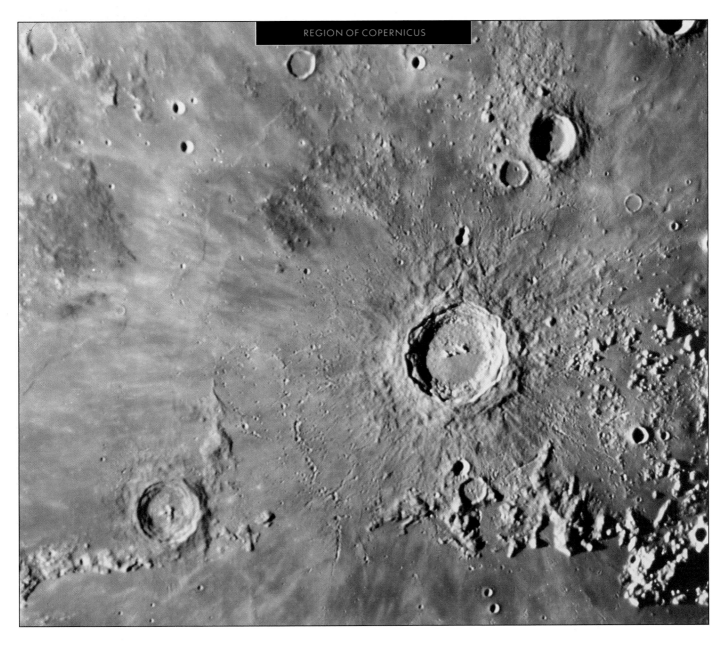

REGION OF COPERNICUS

Named after the Polish astronomer Nicolaus Copernicus, the bright crater at centre of picture (see *above*) is one of the most famous features on the lunar surface. Measuring some 93 km (58 miles) across, its rim reaches to heights of nearly 4 km (2½ miles). The inner walls are terraced while the floor is relatively flat. There are central mountain peaks reaching up to a height of 1·2 km (¾ mile). Much detail can be seen in Copernicus, even with a small telescope. Copernicus, like Tycho (page 24), is

famous for its system of rays which reach out from this young crater. These rays, however, are barely visible when the Sun's illumination is low as when this photograph was taken.

To the lower left (south-west) of Copernicus is the 48-km (30-mile) diameter crater Reinhold, while situated a little further away from Copernicus, towards upper right (north east) of the picture is the prominent crater Eratosthenes. This feature measures 58 km (36 miles) in diameter. Eratosthenes, together with

the other craters seen here, are highlighted by the Sun's rays, emerging from the right of the picture. A little way to the lower left (south-west) of Eratosthenes are the circular remains of the crater Stadius. This 64-km (40-mile) diameter feature has walls reaching to heights of just 650 m (2,100 ft) above the surrounding floor of Sinus Aestuum (Bay of Billows). Stadius is a submerged crater which is extremely difficult to see unless the illumination from the Sun highlights its otherwise undistinguished walls.

PLANETARY DATA

EQUATORIAL DIAMETER (km/miles) :	4,878/3,031
MASS (EARTH = 1) :	0·055
VOLUME (EARTH = 1) :	0·06
AXIAL ROTATION PERIOD (days) :	58·65
AXIAL TILT (°) :	0·0
ORBITAL PERIOD (days) :	87·97
AVERAGE DISTANCE FROM SUN (km/miles) :	57,910,000/35,990,000
INCLINATION OF ORBIT TO ECLIPTIC (°) :	7·0
MEAN DENSITY (g/cm³) :	5·43
NUMBER OF SATELLITES :	0

EARLY OBSERVATIONS

Before the Space Age the only information we had about Mercury was gleaned from Earth-based observations. In the 1880s, the Italian astronomer Giovanni Virginio Schiaparelli (pictured *right*) put together the first chart of the Mercurian surface. His chart showed only bright and dark areas.

Schiaparelli's work was followed up between 1924 and 1933 by the French astronomer Eugenios Antoniadi, although his efforts have since been found to be inaccurate with little correlation between his chart and those drawn up following the visits to Mercury by the Mariner 10 probe (see below). Antoniadi, like Schiaparelli before him, thought that Mercury had a captured rotation and

that the Mercurian day and year were of the same 88-day duration. Had this been the case, the planet would have kept the same, permanently-illuminated face turned towards the Sun, the opposite hemisphere experiencing everlasting night. However, measurements carried out during the early 1960s revealed Mercury's axial rotation period to be 58·65 days. Comparison with the orbital period shows that Mercury rotates three times on its axis during two orbits of the Sun and that the whole of Mercury's surface receives sunlight at one time or another. Another effect of this so-called spin-orbit coupling, and one that led Shiaparelli and Antoniadi to their erroneous conclusion, is that Mercury presents the same face to Earth each time it is best placed for observation.

SCHIAPARELLI (1835–1910)

THE MERCURIAN SURFACE

On 29 March, 1974 the American space probe, Mariner 10, made the first ever fly-by of Mercury. This was followed by further passes in September 1974 and March 1975. The images received during this time have allowed us to accurately map almost half the Mercurian surface. During the first encounter the Mariner 10 cameras provided this view (see *left*), compiled from 18 separate photographs taken from a range of just over 200,000 km (125,000 miles).

The Mariner 10 cameras detected a lunar-type landscape with huge numbers of craters visible. These range in size from several hundred kilometres down to the smallest features seen, which have diameters of 100 metres (330 ft) or so, the limit of resolution of the Mariner 10 cameras. Mountains and valleys were also found to be present, together with long meandering cliffs stretching out over the surface. Yet there were very few of the dark, maria-type features found on the Moon. The chief Mercurian basin, seen here emerging from the day-night terminator at left centre, has been named the Caloris Basin.

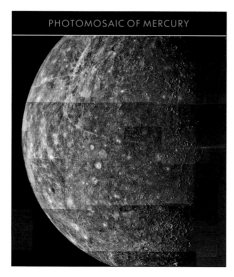

PHOTOMOSAIC OF MERCURY

MERCURY'S CORE

Mercury is thought to have an iron core. Partly molten, this has a diameter approaching 4,000 km (2,500 miles) and occupies over three-quarters of Mercury's total size and nearly half its volume! This is overlaid by a solid mantle and rocky crust which together reach a depth of around 600 km (370 miles).

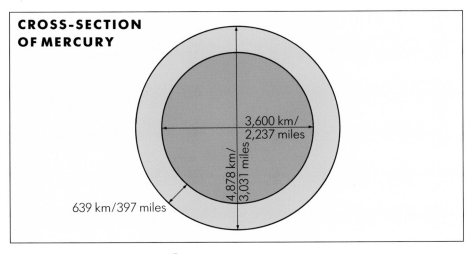

CROSS-SECTION OF MERCURY

3,600 km/ 2,237 miles

4,878 km/ 3,031 miles

639 km/397 miles

CRATER KUIPER

The bright crater seen here (*below*) on the rim of a larger, much older crater has been named Kuiper, in honour of the American scientist, Gerard P Kuiper. A member of the Mariner 10 mission team, Kuiper died in December 1973 while the spacecraft was en route to Mercury. This was the first crater to be identified from Mariner 10 images. It has a diameter of 41 km (25 miles) and is so bright that it was visible in images sent back while Mariner 10 was still more than 3·2 million km (2 million miles) from the planet. The larger crater, Murasaki, is some 80 km (50 miles) across. This image was obtained from a range of 88,450 km (55,000 miles) around 2½ hours before the probe's closest approach.

MERCURIAN CRATERS, KUIPER AND MURASAKI

VENUS

PLANETARY DATA

EQUATORIAL DIAMETER (km/miles)	: 12,102/7,520
MASS (EARTH = 1)	: 0·82
VOLUME (EARTH = 1)	: 0·86
AXIAL ROTATION PERIOD (days)	: 243·01
AXIAL TILT (°)	: 177·34
ORBITAL PERIOD (days)	: 224·7
AVERAGE DISTANCE FROM SUN (km/miles)	: 108,200,000/67,232,000
INCLINATION OF ORBIT TO ECLIPTIC (°)	: 3·4
MEAN DENSITY (g/cm³)	: 5·25
NUMBER OF SATELLITES	: 0

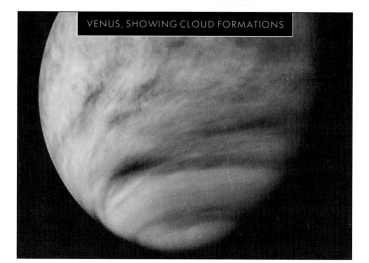
VENUS, SHOWING CLOUD FORMATIONS

OBSERVATION

Venus, like Mercury, orbits the Sun inside the Earth, although its maximum elongation is much greater than that of Mercury. The accompanying diagrams (*right*) show the orbits of the two inner planets together with their apparent positions in the sky in relation to the Sun. As we can see, both Mercury and Venus are best seen either in the east before sunrise or in the west after sunset. Venus is actually so bright it can sometimes be seen in broad daylight, if one knows exactly where to look. However, bright though Venus is, there is little the naked-eye observer can do other than watch the planet as it moves through the sky.

Binoculars show very little when observing Venus, as the glare from the planet will distort the image. When observed against a bright sky, powerful binoculars may bring out the Venusian phases. Remember, however, that casually sweeping the daylight sky with binoculars when the Sun is above the horizon is dangerous and must **never** be attempted.

Venus, like Mercury, displays phases, although these are not generally visible without a telescope. Virtually any small telescope will show the Venusian phases although even very large telescopes will reveal only the top of the dense Venusian atmosphere. Some markings and/or brighter patches may be visible on the cloud deck, although usually only to experienced observers.

ORBITS OF MERCURY AND VENUS

Earth

plane of the planet orbits

East   South   West

Sun rising   East

West   Sun setting

VENUSIAN CLOUDS

Venus is the brightest of the planets, its brilliance being due to the dense clouds that completely cover it (see *above*), which, consisting mainly of carbon dioxide (CO_2), are responsible for the very high surface temperature. CO_2 is a 'greenhouse' gas; its presence in the Venusian atmosphere allows heat received from the Sun to reach the surface, but prevents it escaping back into space. Pollution has produced rises in the level of CO_2 in our planet's atmosphere. A slow rise in temperatures is producing 'global warming', a direct effect of the CO_2 in the atmosphere. So great are the amounts of CO_2 in the Venusian atmosphere, however, that a runaway greenhouse effect has occurred, raising the temperatures at the surface to a staggering 450°C (840°F)! It is not known whether the severity of this effect has always existed.

It is not only the high surface temperatures that make Venus an inhospitable planet. Around 90 per cent of the 250-km (155-mile) deep Venusian atmosphere lies within 28 km (17 miles) of the surface, producing a surface atmospheric pressure 90 times that at the Earth's surface. The Venusian atmosphere is also highly corrosive, containing other substances such as hydrogen sulphide, carbon monoxide and sulphur dioxide together with traces of hydrochloric and hydrofluoric acids.

SPACE-PROBE INVESTIGATION

Needless to say, our knowledge of Venus was somewhat scant until fairly recently, the dense cloud cover entirely blotting out the surface from view of Earth-based telescopes. It is only since space probes have visited the planet that we have started to gather information on this inhospitable world. Probes from the Soviet Venera series gave us our first views of the Venusian surface, the first picture, a single black and white image, being transmitted back from Venera 9 following its landing in October 1975. The first colour pictures of the surface were transmitted by Venera 13 in March 1982.

Landers are only able to provide information on the area on which they set down. In order to study the planet as a whole, different methods are required. The American Pioneer Venus 1 mission in 1978 provided the first radar maps of the Venusian surface. The radar image shown here (*below right*) was taken during the Soviet Venera 15 and 16 missions in 1983 and reveals an elliptical crater measuring around 50×70 km (31×43 miles). It was probably formed during a meteorite impact. Resolution of this image is about 1·5 km (1 mile).

Both the Soviet and American radar mapping missions have enabled scientists to chart most of the Venusian surface. The principle behind this technique is that radar waves can pass through clouds although they are reflected by solid surfaces. Radar mappers are put into orbit around the planet, following which radar pulses are transmitted down to the surface from where they are reflected back to the orbiting craft. The shorter the distance that the signal has to travel, the quicker its return to the orbiter. In other words, pulses hitting mountainous regions came back quicker than those which hit valleys and other lowland features.

The radar mappers have shown that most of the Venusian surface consists of flat plains. There are several highland regions, including Ishtar Terra, which is comparable in size to Australia, and Aphrodite Terra, similar in extent to Africa.

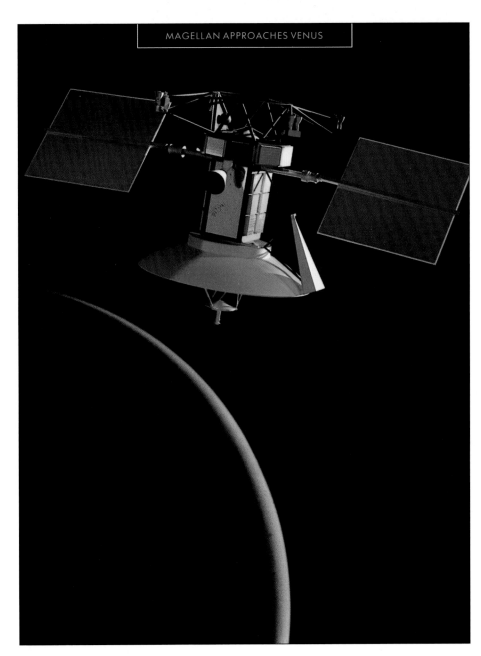

MAGELLAN APPROACHES VENUS

Smaller highland regions have been detected together with numerous mountains, craters, valleys and possibly active volcanoes.

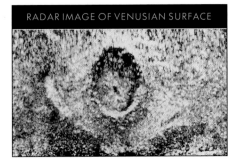

RADAR IMAGE OF VENUSIAN SURFACE

Following in the tracks of Pioneer Venus 1 and Venera 15 and 16 is the American Magellan mission. Launched from the Space Shuttle during Shuttle mission STS-41 in May 1989, Magellan entered into orbit around Venus in August 1990, as shown in this artist's impression (see *above*). The probe will map between 70 and 80 per cent of the Venusian surface during 1,790 orbits of the planet and will resolve features down to 120 metres (394 ft) across, a resolution far better than any previous radar-mapping mission. Magellan promises to revolutionize our knowledge of Venus.

MARS

DIAMETER (km/miles) :	6,786/4,217
MASS (EARTH = 1) :	0·107
VOLUME (EARTH = 1) :	0·15
AXIAL ROTATION PERIOD (days) :	24·62
AXIAL TILT (°) :	25·19
ORBITAL PERIOD (days) :	686·98
AVERAGE DISTANCE FROM SUN (km/miles) :	227,940,000/141,635,000
INCLINATION OF ORBIT TO ECLIPTIC (°) :	1·85
MEAN DENSITY (g/cm³) :	3·95
NUMBER OF SATELLITES :	2

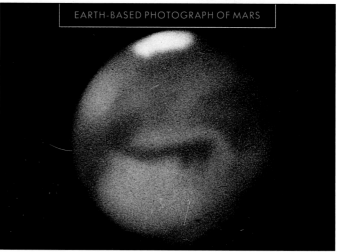

EARTH-BASED PHOTOGRAPH OF MARS

SATELLITE DATA

NAME	DIAMETER (km/miles)	DISTANCE FROM CENTRE OF PLANET (km/miles)	ORBITAL PERIOD (days)	MEAN DENSITY (GM/CM³)
PHOBOS	28 × 20/17 × 12	9,380/5,830	0·319	2·0
DEIMOS	16 × 12/10 × 7·5	23,460/14,580	1·263	1·7

THE MARTIAN LANDSCAPE

This Earth-based telescopic view (see *above right*) highlights the reddish colour of Mars, which is caused by reddish dust scattered over the Martian surface giving rise to the popular name, the Red Planet. Various surface markings can also be seen with one of the two polar ice caps, the latter first being observed in the seventeenth century by the Italian astronomer Giovanni Domenico Cassini. Although not visible here, other Martian features have been observed by Earth-based astronomers, including craters, first noted by the American astronomer Edward Emerson Barnard around a century ago. Barnard's observations were substantiated when space probes visited the planet, their cameras revealing large numbers of craters on the Martian surface.

MARTIAN LINEAR FEATURES

It is only comparatively recently that the existence of Martian life has been cast into doubt after many years of speculation. In 1877 the Italian astronomer Giovanni Schiaparelli observed and recorded a series of linear features on the Martian surface (see diagram). He referred to these features as *canali*, a word signifying natural channels, and made no indication that he believed them to be artificial. However, several astronomers – notably the American Percival Lowell – mistranslated the word as 'canals', a type of feature that is anything but natural. Lowell, among others, came to the conclusion that a Martian civilization had built the canals as part of a huge irrigation system whereby water was carried from the Martian polar ice caps to irrigate vegetation in the equatorial regions! Many astronomers were sceptical about the existence of Martian

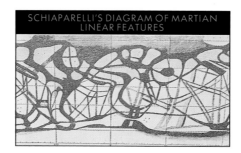

SCHIAPARELLI'S DIAGRAM OF MARTIAN LINEAR FEATURES

canals, although the idea was not finally put to rest until space probes visited the planet. No canals were seen and the supposed canal network seems to have been nothing more than an optical illusion mixed with wishful thinking!

THE MARINER PROBES

The exploration of Mars by space probe started with the launch of the American Mariner 4 craft which passed Mars in 1965. Mariner 4 sent back 21 pictures of the Martian terrain which finally put paid to any ideas of a Martian canal system. Mariners 6 and 7 followed during the late 1960s, their pictures giving us excellent views of the Martian surface, including this photograph (see *opposite top*) taken by Mariner 6 of an area pitted with craters, including the 20-km (12½-mile) diameter feature seen at top of picture.

The Mariner 9 mission greatly expanded our knowledge of Mars. Launched on 30 May, 1971, Mariner 9 entered the orbit of Mars on 14 November of that year, transmitting well over 7,000 pictures of both Mars and its satellites before contact was lost in October 1972.

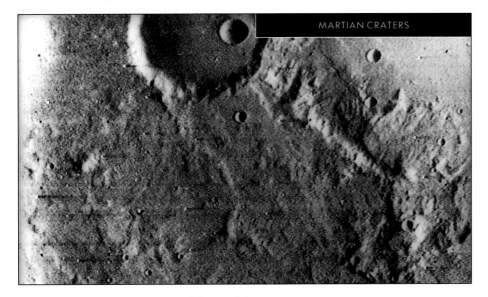

MARTIAN CRATERS

THE VIKING MISSIONS

The two American Viking probes were launched in August (Viking 1) and September (Viking 2) 1975 and arrived at Mars in 1976, becoming the first successful Mars soft-landers. Each Viking probe consisted of an orbiter and lander. Between them, the orbiters returned over 55,000 images of the Martian terrain, from which scientists were able to map the entire planet. These images provided many stunning views, including the photograph of the Valles Marineris (see *below left*), a huge and extensive network of valleys stretching 4,000 km (2,500 miles) across the Martian surface. Viking 1 landed at around latitude 22°N in Chryse Planitia, touching down in a region which, as shown here (see *below right*), consists of an undulating plain strewn with rocks. Viking 2 set down on the opposite side of the planet, at latitude 48°N in Utopia Planitia.

The Viking landers transmitted much information including data on the Martian weather, wind direction and velocity, and information on the lower Martian atmosphere, which was found to consist primarily of carbon dioxide and to have a pressure of less than 1 per cent that at the Earth's surface. The Viking landers also carried out tests for signs of life, each of which involved the examination of soil samples gathered up by remote-control sampler arms. The results of these experiments were not conclusive, but reduced, without eliminating entirely, the possibility that life exists on Mars. It seems, however, that final conclusions will be drawn only after the launch of return-sample missions to the Red Planet.

VALLES MARINERIS

CHRYSE PLANITIA

THE GAS GIANTS: JUPITER

PLANETARY DATA

EQUATORIAL DIAMETER (km/miles)	: 142,984/88,846
MASS (EARTH = 1)	: 317·83
VOLUME (EARTH = 1)	: 1323
AXIAL ROTATION PERIOD (hours)	: 9·84
AXIAL TILT (°)	: 3·12
ORBITAL PERIOD (years)	: 11·86
AVERAGE DISTANCE FROM SUN (km/miles)	: 778,330,000/483,632,000
INCLINATION OF ORBIT TO ECLIPTIC (°)	: 1·31
MEAN DENSITY (g/cm³)	: 1·33
NUMBER OF SATELLITES	: 16

JUPITER AS SEEN THROUGH SMALL TELESCOPE

SATELLITE DATA

NAME	DIAMETER (km/miles)	DISTANCE FROM CENTRE OF PLANET (km/miles)	ORBITAL PERIOD (days)
METIS	40/25	127,960/79,510	0·295
ADRASTEA	24 × 16/15×10	128,980/80,140	0·298
AMALTHEA	270 × 150/168 × 93	181,300/112,700	0·498
THEBE	100/62	221,900/137,900	0·675
IO	3,630/2,256	421,600/262,000	1·769
EUROPA	3,138/1,950	670,900/417,000	3·551
GANYMEDE	5,262/3,270	1,070,000/665,000	7·155
CALLISTO	4,800/2,983	1,883,000/1,170,000	16·689
LEDA	16/10	11,094,000/6,893,000	238·72
HIMALIA	180/112	11,480,000/7,133,000	250·57
LYSITHEA	40/25	11,720,000/7,282,000	259·22
ELARA	80/50	11,737,000/7,293,000	259·65
ANANKE	30/19	21,200,000/13,200,000	631
CARME	44/27	22,600,000/14,000,000	692
PASIPHAE	70/43	23,500,000/14,600,000	735
SINOPE	40/25	23,700,000/14,700,000	758

Discovered by Galileo in 1609-10, and known as the Galilean satellites, the four largest Jovian moons are Ganymede, Callisto, Io and Europa. Both Io and Europa are similar in diameter to our own Moon, while Ganymede and Callisto rival Mercury in size. The Galilean satellites, together with the four other inner Jovian moons, were probably formed in the Jovian vicinity, the outer satellites almost certainly being captured asteroids.

In 1979, the Voyager cameras revealed that Ganymede has two contrasting types of surface, the cratered

regions being much the darker of the two. The lighter regions of grooved terrain, seen between the cratered regions are very different. The comparative lack of cratering in these areas leads to the conclusion that they are much younger than the cratered regions.

Even more heavily cratered than Ganymede is Callisto, one of the most heavily cratered worlds, known. So irregular is the Callistan surface that the only smooth areas detected by the Voyager cameras are those seen at the centres of some of the large impact features on Callisto.

JOVIAN FEATURES

Jupiter displays a prominent polar flattening as can be seen in this image (*above*) which shows Jupiter as it would be seen through a small telescope. The polar flattening is produced through centrifugal forces set up by the rapid axial rotation period of this colossal world. Jupiter also displays numerous bright zones and dark belts, the two prominent equatorial belts clearly seen here straddling the Jovian equator.

CHANGES IN JOVIAN ATMOSPHERE

Jupiter's atmosphere is subject to continuous change, as is evident in this pair of images (see *opposite top*) taken four months apart by Voyager 1 on 24 January, 1979 (left) from 40 million km (25 million miles), and Voyager 2 on 9 May, 1979 (right) from 46·3 million km (29 million miles). Among the changes seen to have taken place were those in the region of the Great Red Spot, notably a white oval feature located just to the south-west (lower left) of the Great Red Spot in the Voyager 1 image. The Voyager 2 picture shows that this oval had drifted 60° to the east (right) in the intervening four months. Many other changes can also be readily seen.

In both these views the Great Red Spot is prominent – a huge atmospheric feature, first glimpsed by the Italian astronomer Giovanni Cassini in 1665. Observed almost continuously ever since, the Great Red Spot is believed to be a storm in the Jovian atmosphere. The colourful clouds we see when we look down on the planet are thought to be contained within the upper 100 km (62 miles) or so of Jupiter's 1,000 km (620 miles) thick atmosphere. Reaching a depth of 20,000 km (12,400 miles) from the base of this atmosphere is a layer of molecular hydrogen, each hydrogen atom having a single electron orbiting a single proton. Beneath this is a 40,000 km (24,900 mile) deep layer of liquid metallic hydrogen; the pressure bearing down upon it is so great that the hydrogen atoms are stripped of their electrons. This produces a soup of protons and electrons, where the atoms are able to move around independently of each other. A solid core, believed to consist of iron and silicates, lies at the Jovian centre.

CHANGES IN JOVIAN ATMOSPHERE

THE JOVIAN RINGS

Jupiter has a ring system, first detected by Voyager 1, measuring 30 km (19 miles) or so thick. The ring system is thought to consist of tiny, rocky particles, the main section of the rings being around 6,400 km (4,000 miles) wide. This lies outside a much fainter halo 20,000 km (12,400 miles) wide which reaches down to within 30,000 km (18,600 miles) of the Jovian cloudtops. Outside these two components is a fainter band extending to around 100,000 km (62,000 miles) from Jupiter.

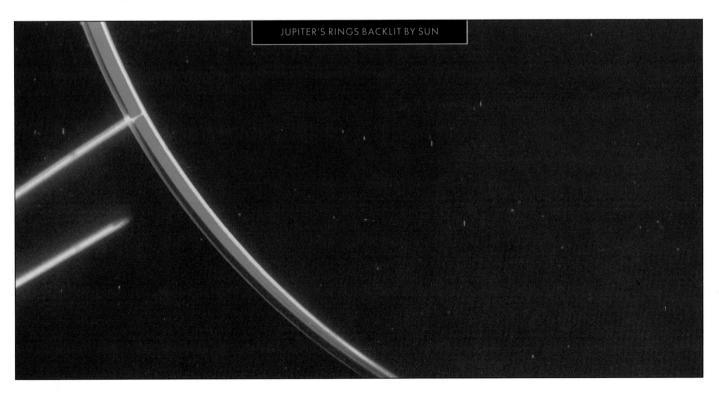

JUPITER'S RINGS BACKLIT BY SUN

IO

This Voyager 1 image (see *below*), taken from a distance of 124,000 km (77,000 miles), shows one of the many volcanic centres discovered on the surface of this Galilean satellite. No impact features were detected on Io. Instead, the surface was found to be covered by sulphur compounds due to frequent volcanic eruptions. These eruptions in turn give Io its bright appearance, the colours displayed by this exciting world including white, yellow, orange and black. The black spot near the lower right of this picture is a volcanic centre, from which long, winding lava flows are seen to emanate.

VOLCANIC CENTRE ON IO

EUROPA

The smallest of the Galilean satellites is Europa which, in appearance, offers stark contrast to Io.

The surface of Europa is smooth, and was found to contain hardly any craters and no mountainous regions. What the Europan surface does contain is a layer of water-ice some 100 km (62 miles) thick, with numerous streaks and cracks, evident in this Voyager 2 image (see *below*) taken from a range of 240,000 km (150,000 miles) These strange features give this billiard-ball of a satellite a fractured appearance.

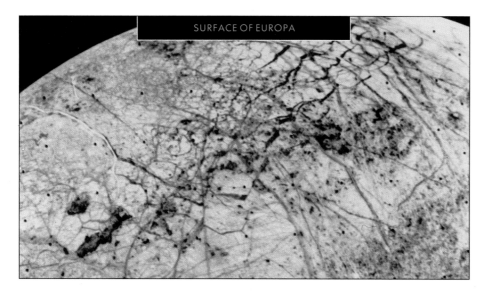

SURFACE OF EUROPA

PLANETARY DATA

DIAMETER (km/miles)	120,536/74,898
MASS (EARTH = 1)	95·181
VOLUME (EARTH = 1)	752
AXIAL ROTATION PERIOD (hours)	10·233
AXIAL TILT (°)	26·73
ORBITAL PERIOD (years)	29·46
AVERAGE DISTANCE FROM SUN (km/miles)	1,426,980,000/886,684,000
INCLINATION OF ORBIT TO ECLIPTIC (°)	2·49
MEAN DENSITY (g/cm³)	0·69
NUMBER OF SATELLITES	18

COLOUR-ENHANCED IMAGE OF SATURN

SATELLITE DATA

NAME	DIAMETER (km/miles)	DISTANCE FROM CENTRE OF PLANET (km/miles)	ORBITAL PERIOD (days)
1981 S13	20/12	133,570/83,000	
ATLAS	40 × 30/25 × 19	137,640/85,530	0·602
PROMETHEUS	140 × 80/87 × 50	139,350/86,590	0·613
PANDORA	110 × 70/68 × 43	141,700/88,050	0·629
EPIMETHEUS	140 × 100/87 × 62	151,422/94,090	0·694
JANUS	220 × 160/137 × 99	151,472/94,120	0·695
MIMAS	390/242	185,520/115,280	0·942
ENCELADUS	500/311	238,020/147,900	1·370
TETHYS	1,050/653	294,660/183,090	1·888
TELESTO	24/15	294,660/183,090	1·888
CALYPSO	30 × 20/19 × 12	294,660/183,090	1·888
DIONE	1,120/696	377,400/234,500	2·737
HELENE	36 × 30/22 × 19	377,400/234,500	2·737
RHEA	1,530/951	527,040/327,500	4·518
TITAN	5,150/3,200	1,221,850/759,220	15·945
HYPERION	350 × 200/217 × 124	1,481,000/920,250	21·277
IAPETUS	1,440/895	3,561,300/2,212,900	79·331
PHOEBE	220/137	12,952,000/8,048,000	550·480

To date, 18 satellites have been found to orbit Saturn, the most recent discovery being that of 1981 S13. This tiny moon, which travels around Saturn within Encke's Division, was spotted by scientists in July 1990 on photographs taken by the Voyager 2 probe around a decade earlier. The largest Saturnian satellite is Titan, a mysterious world found by Voyager 1 to be completely covered in clouds which hide its surface from view. Titan's atmosphere is made up predominantly of nitrogen (90 per cent), with methane and argon being the next most abundant gases. The surface pressure on Titan is 60 per cent greater than that of Earth, and the surface temperature is −177°C (−287°F). At the moment, we can only speculate as to what conditions are like on Titan, although the visit of the ESA/NASA Cassini mission to Saturn later this century should provide some answers. The Cassini mission involves both an orbiter and lander, the orbiter carrying out a survey of the Saturnian system and the lander making a controlled descent through Titan's atmosphere.

SATURN'S COMPOSITION

This image of Saturn (see *above*), obtained by Voyager 1 on 18 October, 1980, has been colour-enhanced to bring out details in its atmosphere. Saturn's polar flattening, like that of Jupiter caused by the planet's rapid axial spin, is also clearly seen.

Beneath the outer visible cloud layers, Saturn is similar in many ways to Jupiter. Both planets have a solid rocky core, although that of Saturn is somewhat larger, being roughly equal in size to the Earth. However, in spite of its somewhat larger core, Saturn's overall density is considerably less than that of Jupiter, or indeed of any other planet. This means that its density must decrease rapidly between the core and outer surface. Indeed, Saturn's layer of liquid metallic hydrogen, located above its core, is around 21,000 km (13,000 miles) deep, only half the depth of that within Jupiter. The overlying, less dense molecular hydrogen layer, however is far deeper than Jupiter's.

SATURN'S NORTHERN HEMISPHERE

When we look at Saturn, we see dark belts and brighter zones similar to those observed on Jupiter, although those of Saturn are less pronounced and nowhere near as colourful. The photograph shows Saturn's northern hemisphere (see *right*). Taken by Voyager 2 from a distance of 7 million km (4·3 million miles) on 19 August, 1981, it has been computer-enhanced to bring out subtle details.

The dark oval feature, seen near top centre of the picture, and the two whitish spots below it are eddies which are moving in a westerly direction at around 65 km (40 miles) per hour. Features that have been observed within the blue band immediately above these eddies have been seen to be travelling in the opposite direction at some 480 km (300 miles) per hour. Determination of wind speeds in Saturn's upper atmosphere have been calculated by tracking features such as those described and shown here. Elsewhere on the planet, much higher wind speeds have been detected, accelerating to as much as 1,600 km (1,000 miles) per hour near the equator.

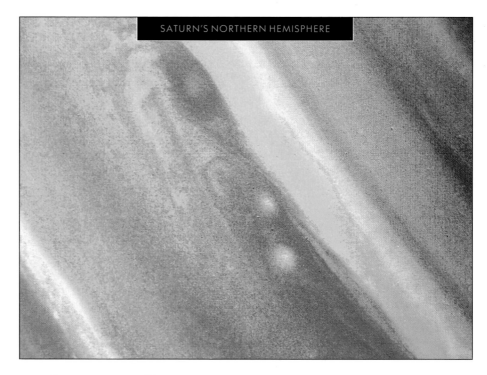

SATURN'S NORTHERN HEMISPHERE

SATURN'S RINGS

Saturn's ring system, regarded by many as the most beautiful sight in the heavens, is made up of countless tiny icy particles and is divided into several sections. The brightest of these is the B-ring, seen here (below) in this Voyager 2 photograph obtained on 22 August, 1981 from a range of 4 million km (2·5 million miles).

Less conspicuous is the A-Ring which is located outside the B-ring, while fainter still is the inner C-ring, or Crepe Ring. All these rings are visible through Earth-based telescopes. Just beyond the A-

ring is the F-ring, first brought to light by Pioneer 11 in 1979, while tucked between the C-ring and the Saturnian cloud tops is the D-ring. Like the faint E- and G-rings, themselves found well beyond the main ring system, the D-ring was detected by the Voyager cameras during their visits to Saturn in 1980 and 1981. Close examination of the rings reveals that they are comprised of many thousands of individual ringlets (see photograph). Also seen here are numerous radial 'spokes'. These are formed from microscopically-small,

dark dust-particles suspended above the main ring plane.

In 1675, the Italian-born French astronomer Giovanni Domenico Cassini discovered a gap between the A- and B-rings, which was subsequently named in his honour. Another division, this time in the outer regions of the A-ring, was spotted by the German astronomer Johann Franz Encke in 1837. The Cassini and Encke Divisions are the two most prominent of the divisions within the Saturnian ring system.

MIMAS

Mimas is the innermost of Saturn's larger satellites. The Voyager cameras have shown that its surface is covered with craters, by far the largest of which is Herschel. This huge impact feature (not visible here) has a diameter of 130 km (80 miles), around a third of the diameter of Mimas itself. The collision that produced Herschel must have come close to shattering the satellite. This Voyager 1 image (see *opposite top*), taken from a range of 129,000 km (80,000 miles) on 12th November, 1980, shows features down to around 2 km (1¼ miles) across.

SATURN'S RING SYSTEM

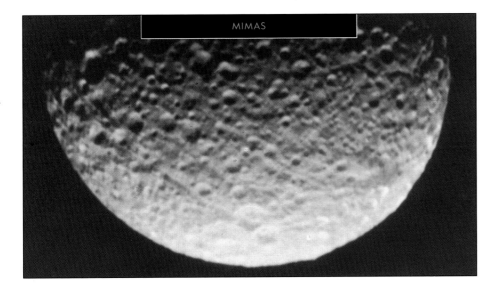

MIMAS

ENCELADUS

This image of Enceladus (*below*), obtained by Voyager 2 from a range of 120,000 km (75,000 miles) on 25 August, 1981 shows numerous ice flows and cracks running across its surface. The icy surface of Enceladus reflects practically all the light received from the Sun back into space, making this satellite the most reflective object in the Solar System. The absence of cratering in the areas dominated by the grooves and flows suggests that the surface here is fairly young. Craters can be seen on Enceladus, some of which are visible on this photograph near the day/night terminator, their forms emphasized by the rays of the setting Sun. These craters appear fresh and, like the nearby plains, seem to be fairly young. The differences in crater frequency suggests that material from inside the satellite has welled up to spread across the surface, covering up features.

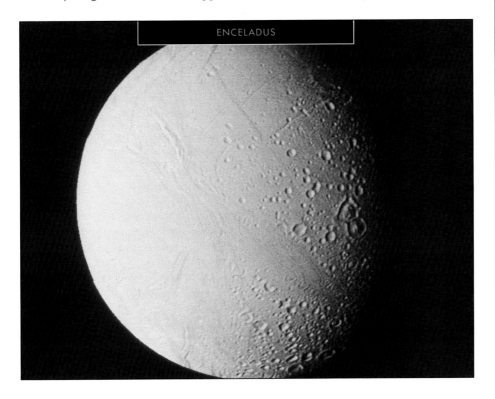

ENCELADUS

THE GAS GIANTS: URANUS

PLANETARY DATA

DIAMETER (km/miles)	51,118/31,763
MASS (EARTH = 1)	14·53
VOLUME (EARTH = 1)	64
AXIAL ROTATION PERIOD (hours)	17·9
AXIAL TILT (°)	97·86
ORBITAL PERIOD (years)	84·01
AVERAGE DISTANCE FROM SUN (km/miles)	2,870,990,000/1,783,950,000
INCLINATION OF ORBIT TO ECLIPTIC (°)	0·77
MEAN DENSITY (g/cm³)	1·29
NUMBER OF SATELLITES	15

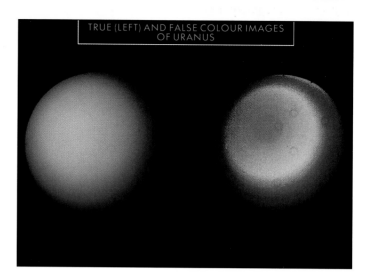

TRUE (LEFT) AND FALSE COLOUR IMAGES OF URANUS

SATELLITE DATA

NAME	DIAMETER (km/miles)	DISTANCE FROM CENTRE OF PLANET (km/miles)	ORBITAL PERIOD (days)
CORDELIA	30/19	49,750/30,910	0·335
OPHELIA	30/19	53,760/33,410	0·376
BIANCA	40/25	59,160/36,760	0·435
CRESSIDA	70/43	61,770/38,380	0·464
DESDEMONA	60/37	62,660/38,940	0·474
JULIET	80/50	64,360/40,000	0·493
PORTIA	110/68	66,100/41,070	0·513
ROSALIND	60/37	69,930/43,450	0·558
BELINDA	70/43	75,260/46,760	0·624
PUCK	150/93	86,010/53,440	0·762
MIRANDA	470/292	129,780/80,640	1·414
ARIEL	1,160/721	191,240/118,830	2·520
UMBRIEL	1,170/727	265,970/165,270	4·144
TITANIA	1,580/982	435,840/270,820	8·706
OBERON	1,520/945	582,600/362,010	13·463

axial tilt, amounting to almost 98°, means that its northern and southern hemispheres point alternately towards the Sun during the planet's 84-year orbit. When this picture was taken, the northern hemisphere was in darkness.

The image at right has been computer processed to dramatically increase contrast, thereby making apparent subtle details in the atmosphere. Visible here are bands girdling the southern (facing) hemisphere. These bands run around well-defined zones and culminate in a much darker hood covering the pole. It has been suggested that the Uranian atmosphere is covered by a haze which, through the planet's axial rotation, has been drawn out into the polar hood and zones seen in this picture.

THE DISCOVERY OF URANUS

Uranus holds the distinction of being the first planet to be discovered telescopically. The discovery was made in March 1781 by William Herschel (see *inset*) who was observing stars in Gemini at the time and stumbled across an object which he at first thought to be a comet. Later observation showed it to be a new planet, orbiting the Sun at roughly twice the distance of Saturn.

URANIAN ATMOSPHERE

Uranus is another of the gas giants, consisting primarily of hydrogen and helium. This photograph (*above right*) was taken by Voyager 2 in 1986 from a distance of just over 9 million km (5·5 million miles). It shows the planet as a blue-green world as it would appear to the human eye. What we see here is the top of the Uranian atmosphere where the temperature is a frigid −210°C (−346°F). Notice the darkening of the disc at upper right of image. This marks the terminator, or boundary between the day and night hemispheres of the planet. Uranus' large

URANIAN SATELLITES

Voyager observations increased the number of known Uranian satellites from five to 15. Although only basic information, such as approximate diameter and orbital periods, were obtained of the smaller moons, the Voyager cameras picked out surface details on the larger satellites. Craters, cliffs and valleys were predominant throughout the larger satellites. In particular, Oberon was also found to have a lofty mountain which towers around 6 km (4 miles) above the surrounding surface.

By far the most geologically interesting of the satellites is Miranda, seen here (*right*) in a photograph taken by Voyager 2 from a distance of just over 30,000 km (18,650 miles). The surface of Miranda was found to possess different types of terrain.

Older, rugged highlands, containing an abundance of craters offered a stark contrast to much brighter regions which had many ridges and valleys with faults cut across them. The ridges and valleys had been caused by the dropping of fault blocks. One of the cliffs created by this tectonic activity is shown here just to the lower right of centre. This was in fact the largest fault captured by the Voyager cameras, and is around 20 km (12 miles) high!

THE URANIAN RING SYSTEM

Uranus's ring system was discovered in March 1977 when the planet passed in front of a star. The light from the star dimmed a number of times both before and after passing behind the planet. This repeated blinking could only have been caused by the star moving behind a system of rings. The distances of the rings range from the innermost 1986 U2R, lying 37,000 km (23,000 miles) above Uranus, to the outermost Epsilon ring, located just over 51,000 km (31,700 miles) from the planet.

There are eleven individual rings, together with smaller particles scattered throughout the ring system. This Voyager 2 image (*above right*), taken from within Uranus's shadow over 230,000 km (143,000 miles) above the Uranian cloud tops, shows that some of the dust is drawn out into distinct lanes.

SHEPHERD SATELLITES

Cordelia and Ophelia, two of the moons discovered by Voyager 2 and seen in this photograph taken in January 1986 (*right*) were found to be shepherd satellites. They orbit Uranus at either side of the Epsilon ring. The gravitational influences of these two satellites maintain the Epsilon ring, preventing the particles within them from spreading out.

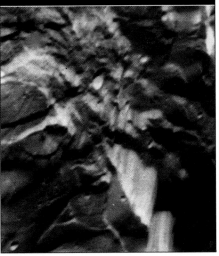

SURFACE OF MIRANDA

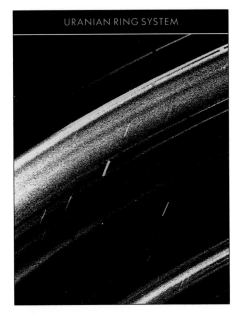

URANIAN RING SYSTEM

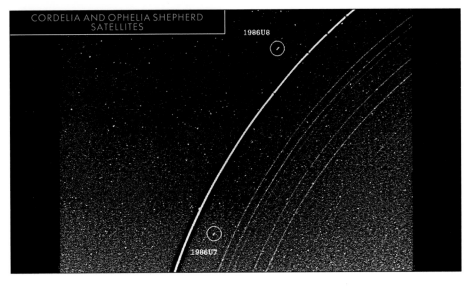

CORDELIA AND OPHELIA SHEPHERD SATELLITES

1986U8

1986U7

NEPTUNE

PLANETARY DATA

EQUATORIAL DIAMETER (km/miles) :	49,528/30,775
MASS (EARTH = 1) :	17·135
VOLUME (EARTH = 1) :	54
AXIAL ROTATION PERIOD (hours) :	19·2
AXIAL TILT (°) :	29·6
ORBITAL PERIOD (years) :	164·79
AVERAGE DISTANCE FROM SUN (km/miles) :	4,497,000,000/2,794,300,000
INCLINATION OF ORBIT TO ECLIPTIC (°) :	1·77
MEAN DENSITY (g/cm³) :	1·64
NUMBER OF SATELLITES :	8

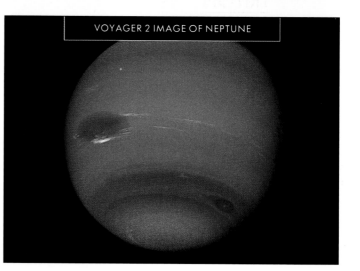

VOYAGER 2 IMAGE OF NEPTUNE

SATELLITE DATA

NAME	DIAMETER (km/miles)	DISTANCE FROM CENTRE OF PLANET (km/miles)	ORBITAL PERIOD (days)
NAIAD	60/37	48,000/29,800	0·296
THALASSA	80/50	50,000/31,000	0·312
DESPINA	150/93	52,500/32,600	0·333
1989 N4	150/93	62,000/38,500	0·429
1989 N2	190/118	73,600/45,700	0·554
PROTEUS	415/258	117,600/73,100	1·121
TRITON	2,700/1,678	354,800/220,500	5·877
NEREID	340/211	5,513,400/3,425,900	360·16

THE DISCOVERY OF NEPTUNE

In the years following its discovery, Uranus was seen to be wandering from its predicted orbit. These deviations suggested to astronomers that another planet may exist beyond Uranus and that it was the gravitational influence of this planet that was tugging Uranus off course. Two mathematicians – the Englishman John Couch Adams and the Frenchman Urbain Jean Joseph Le Verrier – decided to try and calculate where in the sky this hypothetical planet could be found. Eventually, Le Verrier (see *inset*) completed his work and sent his predictions to the Berlin Observatory where, after a short search, Johann Gottfried Galle and Heinrich Louis d'Arrest located the new planet almost exactly where Leverrier had said it would be. Le Verrier's calculations were found to tie in quite well with those of Adams, and now both men are given equal credit for their work.

NEPTUNIAN FEATURES

White methane clouds contrast with Neptune's deep blue atmosphere in this Voyager 2 image (see *above*) taken from a range of 10 million km (6·2 million miles). Even from this distance Neptune's atmosphere was seen to be in motion, changes in some cloud features being evident over just several hours. Particularly prominent here is the Earth-sized Great Dark Spot, which is shown in more detail in the accompanying view (see *right*) obtained from a distance of 2·8 million km (1·7 million miles). This huge dynamic feature, located 22° south of Neptune's equator, is rotating anti-clockwise and is bordered by cirrus-type clouds which were seen to undergo changes in appearance over only a few hours. The smallest clouds seen here are less than 100 km (60 miles) across. The similar Small Dark Spot, can be seen much closer to the Neptune's south pole, at a latitude of around 55°S. Bright clouds forming at the centre of the Small Dark Spot suggest that material is spilling out here from beneath the outer visible surface of Neptune. A cloud feature nicknamed 'The Scooter' is seen here as a triangular patch, but was seen to change as Voyager passed over Neptune. Located at latitude 42°S, it was given its name because it travels around the planet in a much shorter (16·8 hours) period than other bright cloud features.

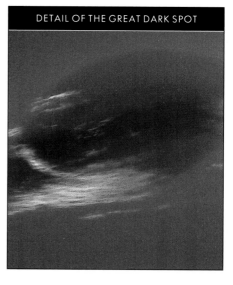

DETAIL OF THE GREAT DARK SPOT

TRITON

The image seen here (*right*) of Triton is actually a composite of about a dozen separate photographs obtained by Voyager during its closest approach to the satellite. The bright surface of the south polar cap seems to be covered by a thin layer of nitrogen and methane frost. Several dark streaks can be seen on the polar cap. These were formed as nitrogen gas erupted from beneath the surface. During these eruptions, dark material was carried up with the gas and lifted to heights of several kilometres (miles) before being spread out across the surface by winds in Triton's tenuous atmosphere. The resulting streaks are up to 150 km (90 miles) long.

Unlike the polar cap, the surface of which may be only a few tens of years old, the cratered terrain seen elsewhere on Triton may be billions of years old. Surface regions away from the polar cap are seen to be criss-crossed with huge ice-filled cracks. The images we have of Triton reveal a varied terrain and indicate a complex geological history. Volcanism certainly played a part in the moulding of the surface we see today and, it appears that Triton is still volcanically active.

Prior to the Voyager encounter, only two satellites, Triton and Nereid, were known to orbit Neptune. The Voyager cameras brought the total to eight, the new discoveries all being small, irregularly shaped worlds. The closest Voyager went to Nereid was 4·7 million km (nearly 3 million miles), and from this range no details were visible on its surface. The satellite Proteus was actually found to be larger than Nereid; it was unknown before the Voyager encounter because it lies much closer to Neptune and is therefore lost in the glare from the planet.

RINGS OF NEPTUNE

Voyager 2 obtained this image (*above right*) of the Neptunian ring system after passing over the planet. Conspicuous here are the outer 1989 N1R ring with the brighter 1989 N2R ring inside it. These rings are located 63,000 km (39,000 miles) and 53,000 km (33,000 miles) from the centre of Neptune. The fact that these two rings are so narrow is probably due to the existence of as-yet undiscovered 'shepherd satellites' orbiting near them. As happens elsewhere in the Solar System, the weak gravitational influence of these satellites maintains the extent of the rings. The dimmer and much wider 1989 N3R ring is also visible, lying around 42,000 km (26,000 miles) from Neptune's centre. Finally, a very tenuous sheet of dust stretches towards the planet from a point between the two outer rings. Designated 1989 N4R, this sheet has a width of just under 6,000 km (3,700 miles). The particles of fine dust that form the ring system are actually darker than coal, although they appear bright as a result of their position between the Sun and the spacecraft. Sunlight is backscattered by the particles in much the same way that dust on a car windscreen is backscattered when the car is facing the Sun.

The references given to the rings are temporary. They are derived from their year of discovery (1989), the initial of the planet (N), the numerical sequence of discovery, and a letter indicating that it is a ring. Newly-discovered satellites, both at Neptune and around other planets, are given similar references, proper names eventually replacing the numerical designations.

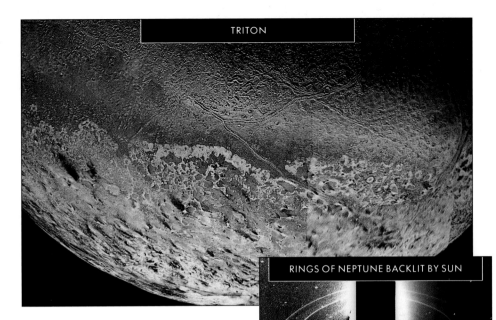

TRITON

RINGS OF NEPTUNE BACKLIT BY SUN

OBSERVATION

Neptune never exceeds 8th magnitude, even when at opposition, and is therefore well below naked-eye visibility.

With the help of a suitable finder chart, binoculars will allow you to locate Neptune against the background sky. Once located, however, the average observer will be able to do nothing more than plot its changing position as the planet travels along its orbit.

Telescopically, Neptune displays a pale blue-green disc, although this will be featureless except when viewed through very large telescopes. Powerful instruments will reveal Neptune's brighter equatorial regions together with traces of belts running across the planetary disc. Its angular diameter is only around 2.2 seconds of arc. At magnitude 13·5, when Neptune is at or near opposition, its largest satellite Triton may be glimpsed in medium-sized telescopes. Of the other satellites, only Nereid is ever visible telescopically, and even then only through a very large telescope.

PLUTO

PLANETARY DATA

DIAMETER (km/miles)	: 2,300/1,430
MASS (EARTH = 1)	: 0·002
VOLUME (EARTH = 1)	: 0·01
AXIAL ROTATION PERIOD (days)	: 6·387
AXIAL TILT (°)	: 122·46
ORBITAL PERIOD (years)	: 248·54
AVERAGE DISTANCE FROM SUN (km/miles)	: 5,913,500,000/3,674,500,000
INCLINATION OF ORBIT TO ECLIPTIC (°)	: 1·77
MEAN DENSITY (g/cm³)	: 2·03
NUMBER OF SATELLITES	: 1

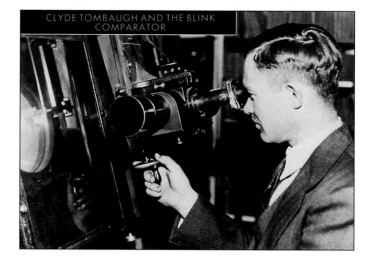

CLYDE TOMBAUGH AND THE BLINK COMPARATOR

SATELLITE DATA

NAME	DIAMETER (km/miles)	DISTANCE FROM CENTRE OF PLANET (km/miles)	ORBITAL PERIOD (days)
CHARON	1,190/740	19,640/12,200	6·387

PERCIVAL LOWELL AND THE HUNT FOR PLUTO

Neptune had been discovered following calculations of its position based on observed perturbations in the orbit of Uranus, and the discovery of Pluto followed a similar path.

The idea of another planet orbiting the Sun beyond Neptune was taken up by several astronomers, notably the American observer Percival Lowell (see *inset*). He calculated the position of a trans-Neptunian planet based on continued discrepancies in the orbit of Uranus. Lowell himself searched for the planet from the Lowell Observatory in Arizona before his death in 1916, although he had no success.

Before he died however, Lowell instigated the construction of a special wide-field camera with which a more thorough search could be made for the trans-Neptunian planet.

THE BLINK COMPARATOR: THE SEARCH CONTINUES

It was not until 1929 that the search continued. Clyde Tombaugh, a young astronomer who had come to the Lowell Observatory to carry out the search, used the special wide-field camera devised by Lowell to try and track down the planet. The basic method behind Tombaugh's search was to examine pairs of photographs taken of selected regions of sky. Each region was photographed twice on photographic plates taken several days apart. Examination of these photographs was carried out using an instrument called a blink comparator, seen here being used by Clyde Tombaugh (see *above*).

Because stars are fixed in space their positions would not change during the interval between each photograph. However any moving object, such as a planet, would appear to shift in relation to the background stars between the times of the two exposures. The plates were put side by side into the blink comparator and then projected alternately onto a screen. Unlike the stars, which

would appear in exactly the same positions on each image, any object which had moved between the times of the two exposures would appear to jump from side to side on the screen.

THE DISCOVERY OF PLUTO

In February 1930, after examining tens of millions of star images, Tombaugh spotted the new planet near the star Delta Geminorum. It appeared as a dim, 15th magnitude point on images taken six days apart during the previous January. These two images can be seen at the top of page 43. The eventual location of the new planet was a truly Herculean task. Some of the plates obtained during the search contained literally hundreds of thousands of star images. Bearing in mind that Pluto's image would appear as nothing more than a faint, starlike point of light and would only be revealed through its motion against the background stars, Tombaugh's efforts are worthy of the highest praise. The planet was named Pluto after the mythological Guardian of the Underworld.

Pluto has an extremely eccentric orbit and, for 20 years out of every 248-year journey around the Sun, it actually comes within the orbit of Neptune. However, Pluto's orbital plane is tilted a full 17° to the ecliptic, thereby removing any chance of a collision between Neptune and Pluto. Pluto last crossed Neptune's orbit in 1989 and will remain closer to the Sun than Neptune until 1999 when it will once more resume its role as the outermost member of the Sun's planetary family.

CHARON

Although Pluto's orbit takes it through the outer regions of the Solar System, it is not on its own as it orbits the Sun. In 1978, the American astronomer James W Christy was examining photographs of Pluto when he spotted an irregularity in Pluto's shape. Christy noticed what appeared to be a bump on Pluto's disc. Examination of other images revealed similar bumps, leading Christy to the conclusion that Pluto had a satellite, which he named Charon, after the ferryman who carried dead souls across the River Styx to Pluto's Underworld. Following its discovery, observation of Charon's orbital motion around Pluto allowed astronomers to calculate its density which, like that of Pluto itself, turns out to be only around twice that of water. In light of this, both Pluto and Charon seem to be made up of a mixture of frozen methane and rocky material.

Their close proximity to each other, coupled with Pluto's distance from Earth, means that Pluto and Charon are extremely difficult to resolve through Earth-based telescopes. The photograph shown here (*right*) includes a ground-based image at top left. However, both objects are resolved in the top right image, obtained with the Hubble Space Telescope. The diagram beneath these two images shows the relative positions of Pluto and Charon at that time ·

Observation of Charon's orbital motion also revealed that its orbital period was equal to the axial rotation period of Pluto itself. Charon permanently hangs above the same point on Pluto's surface and, to a theoretical 'observer' on the Charon-facing hemisphere of Pluto, is forever suspended in the Plutonian sky. Charon also turns out to be around half the size of its parent planet, leading astronomers to regard the system as a double-planet.

PLUTO DISCOVERY PLATE, TAKEN 23 JANUARY, 1930

PLUTO DISCOVERY PLATE, TAKEN 29 JANUARY, 1930

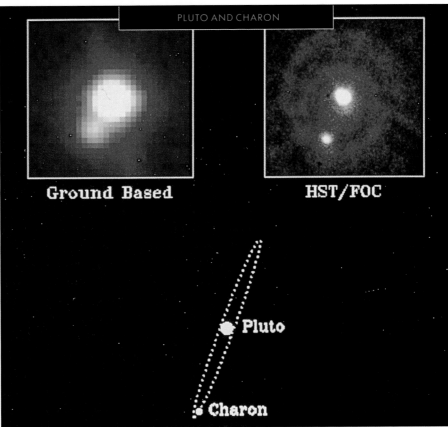

PLUTO AND CHARON

Ground Based

HST/FOC

Pluto

Charon

THE ASTEROIDS

Suggestions as to the origin of the asteroids, or minor planets, have been many. Now regarded as remnants of material which never collected to form a planet, the asteroids are a selection of relatively tiny objects, most of which are found between the orbits of Mars and Jupiter. Even the largest asteroid Ceres, dis-covered by the Italian astronomer Giuseppe Piazzi on 1 January, 1801, is less than 1,000 km (620 miles) across. Although numbering many thousands, the collective mass of the asteroids would produce an object with a diameter considerably less than that of even the smallest of the major planets.

THE FIRST TEN ASTEROIDS

NO	NAME	DISCOVERER	DATE OF DISCOVERY	DIAMETER (km/miles)	ORBITAL PERIOD (years)
1	CERES	PIAZZI	1 JANUARY, 1801	940/584	4·60
2	PALLAS	OLBERS	28 MARCH, 1802	588/365	4·62
3	JUNO	HARDING	1 SEPTEMBER, 1804	248/154	4·36
4	VESTA	OLBERS	29 MARCH, 1807	576/358	3·63
5	ASTRAEA	HENCKE	8 DECEMBER, 1845	120/75	4·13
6	HEBE	HENCKE	1 JULY, 1847	204/127	3·78
7	IRIS	HIND	13 AUGUST, 1847	208/129	3·69
8	FLORA	HIND	18 OCTOBER, 1847	162/101	3·27
9	METIS	GRAHAM	26 APRIL, 1848	158/98	3·69
10	HYGEIA	DE GASPARIS	12 APRIL, 1849	430/267	5·55

THE SEARCH FOR ASTEROIDS

Until the end of the 19th century, aster-oids were discovered by astronomers patiently peering through their tele-scopes. A few observers chalked up sig-nificant numbers of finds including Johann Palisa, who discovered 121, and Karl Reinmuth who leads the way with 246 to his name. Second to Reinmuth is the German astronomer Maximilian

EROS

Among the known examples of Earth-grazing asteroids is Hermes which came within 900,000 km (560,000 miles) of our planet in October 1937. Even this was surpassed by the tiny 300-m (984-ft) diameter object designated 1989FC which approached to within 700,000 km (435,000 miles) of us in March 1989.

A famous Earth-grazing asteroid is the Apollo-type object Eros, discovered by G Witt from Berlin in 1898. This aster-oid is elongated, measuring some 10 × 20 × 30 km (6 × 12 × 18 miles). This unusual shape was first suspected after variations in its magnitude were detected in 1900, these variations being caused by Eros presenting different surfaces towards us. This artist's impression (*below*) shows Eros's orbital path around the Sun in relation to that of the Earth.

PHOTOGRAPHIC TRAIL OF ASTEROID 28 BELLONA

Franz Joseph Cornelius Wolf (better known as Max Wolf). Wolf was success-ful in his hunt for asteroids by using photography. Wolf's method was simple but effective. He photographed selected regions of the sky with a camera attached to a telescope, the telescope being driven to track the stars as they passed across the sky. Each star appeared as a point of light on the photograph, although any asteroids in the same field of view appeared as streaks of light, caused by their motion through the sky.

A typical asteroid trail, that of 28 Bellona, is seen here on a photograph taken in December 1908 (see *above*). The first asteroid to come to light through Wolf's efforts was 323 Brucia, spotted on 20 December, 1891.

ASTEROID ORBITS

Although most asteroids keep to the area between the orbits of Mars and Jupiter (known as the Asteroid Belt), a number have been found to wander from this region (see diagram *below*). These asteroids are members of groups, each of which is comprised of objects that follow particular types of orbit. One of these groups consists of the Earth-crossing Aten asteroids, Aten itself travelling in a path which carries it between 118 and 170 million km (73 and 106 million miles) from the Sun. Another Earth-crossing group are the Apollo asteroids. The distance of Apollo from the Sun varying between 97 and 343 million km (60 and 213 million miles). Although both Aten and Apollo asteroids are Earth-crossing, the groups are distinguished from each other by their orbits. The average dis-tances from the Sun of Aten asteroids are less than that of the Earth, while those of Apollo asteroids are greater.

Trojan asteroids, of which well over 200 have been found, travel around the Sun in orbits that virtually match that of Jupiter. They form two distinct groups, and are named after heroes in the Trojan wars. One group precedes Jupiter in its orbit and the other follows, the angular distance measured from the Sun of the two groups being 60° in each case.

Many other asteroids orbit the Sun in paths that take them away from the main Asteroid Zone, including the Amor asteroids. As with other asteroids in its class, Amor is a Mars-crossing asteroid, its distance from the Sun varying between 162 and 412 million km (101 and 256 million miles).

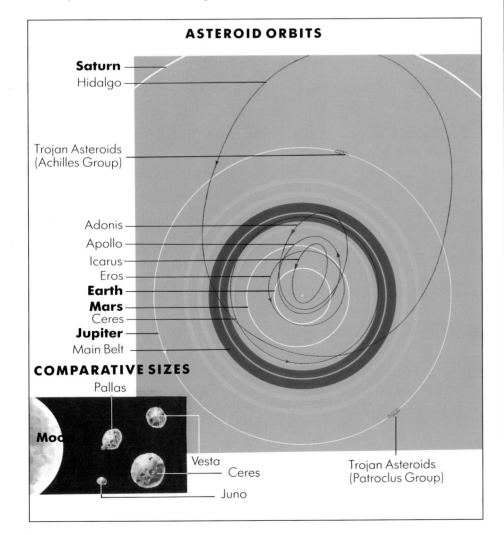

ASTEROID ORBITS

Saturn
Hidalgo
Trojan Asteroids (Achilles Group)
Adonis
Apollo
Icarus
Eros
Earth
Mars
Ceres
Jupiter
Main Belt

COMPARATIVE SIZES

Pallas
Moon
Vesta
Ceres
Juno

Trojan Asteroids (Patroclus Group)

COMETS

Although the orbits of many comets have been accurately plotted, previously unknown comets can appear at any time. Although sometimes visually impressive, most comets are fairly unspectacular and generally observable only with the aid of binoculars and telescopes.

THE ORIGIN OF COMETS

Comets contain relatively little mass, consisting primarily of a frozen mixture of rock and dust. They are thought to originate within Oort's Cloud, a vast swarm of primaeval material orbiting the Sun beyond the orbit of Pluto. The existence of this cloud was first suggested by the Dutch astronomer Jan Oort in 1950 and is believed to consist of icy particles remaining after the formation of the Solar System. Chunks of material are ejected from Oort's Cloud, perhaps by the gravitational influence of passing stars. These chunks then head either away from the Solar System or towards the inner regions of the Sun's family. As a comet draws in towards the Sun, solar energy acts on the comet resulting in the vaporization of ice. This in turn releases dust and gas which then forms a fuzzy head, or 'coma', around what becomes the nucleus of the comet.

THE ORBITS OF COMETS

The lengths of cometary orbits vary, ranging from the shortest known orbital period of 3·3 years, belonging to Encke's Comet, to orbits that are so long we are unable to predict them at all accurately. Comets lose material every time they pass through the inner Solar System, this material being blown away by the relentless force of solar energy. It follows that a comet will become fainter and more difficult to observe with the pas-

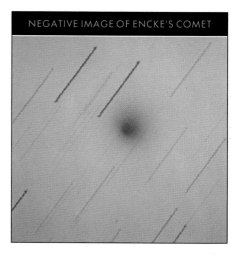
NEGATIVE IMAGE OF ENCKE'S COMET

sage of time. Encke's Comet, seen here (*above*) in this negative image taken on 29 October, 1914, has suffered more than most. Encke's Comet was quite prominent when first discovered two centuries ago.

The material shed by a comet becomes spread out along its orbital path. At certain times of the year the Earth's orbit carries it through the orbital paths of certain comets. When this occurs, a larger number of particles than normal enter the atmosphere producing a sometimes dramatic rise in the number of observed meteors (see Meteors and Meteorites, pages 48-49).

COMETARY TAILS

The material released from the nucleus is blown away by solar energy to form tails. These may consist of dust or gas; in the case of the latter, the matter is ionized (each atom loses one or more electrons). Comets can have more than one tail, such as De Cheseaux' Comet discovered in 1743 which had at least six! Unusual activity and structure can often be seen in cometary tails, such as that of Morehouse's Comet of 1908 seen here (*below*). The tail of this comet separated from the nucleus during early-October, the material from it eventually dissipating into space. A new tail emerged with-in a couple of weeks. Because it is the constant force of solar energy that creates the tail and blows it away from the head of the comet, cometary tails are always seen to point away from the Sun regardless of its direction of travel. As the comet rounds the Sun and heads off back into space, the tails and coma disappear. Eventually the comet becomes an icy nucleus once more. If the comet should wander within the gravitational influence of one of the major planets it may be tugged into a shorter orbit, perhaps ensuring that it may once again grace our skies.

MOREHOUSE'S COMET

NAMING COMETS

Comets normally bear the name of their discoverers, although in some cases they are named after the individual who predicts their return following previous sightings. Such was the case with Halley's Comet, pictured here from Chile, as seen in the evening twilight, in January 1986 (see *below*). The English astronomer Edmund Halley noted that the bright comet he saw in 1682 had an orbit similar to comets seen in 1531 and 1607. He deduced that all sightings were of the same object and that it would return in 1759. It did, and was named in Halley's honour.

HALLEY'S COMET

THE GIOTTO PROBE AND HALLEY'S COMET

The return of Halley's Comet in 1985-6 provided scientists with their first-ever opportunity to explore a comet with space probes. Of the numerous probes sent to examine Halley's Comet, the European Giotto probe was by far the most successful.

Giotto travelled through the coma to within 610 km (380 miles) of the nucleus of Halley's Comet on 14 March, 1986. Its cameras returned images revealing the nucleus to be a tiny and irregularly shaped chunk of ice 15 km (9 miles) long by 8 km (5 miles) wide. Numerous bright spots – jets of dust and gas – can be seen in this image (*right*). Three vents were also found located in the outer dust layer, the material forming the coma and tails of Halley's Comet

COMET IRAS-ARAKI-ALCOCK

The maximum number of names that can be attached to a comet is three. This has happened on several occasions as was the case with Comet IRAS-Araki-Alcock, seen here (*above*) on a photograph taken on 8 May, 1983. This comet was discovered independently by the joint USA/UK/Dutch Infra-red Astronomy Satellite (IRAS), the Japanese observer Genichi Araki, and the English astronomer George Alcock in 1983.

NUCLEUS OF HALLEY'S COMET

emerging through them. In spite of several instruments (including the optical camera) being put out of action following a collision between Giotto and a dust particle from the comet, Giotto emerged from the encounter in reasonable working order, and has now been retargeted towards an encounter with Comet Grigg-Skjellerup in July 1992.

METEORS AND METEORITES

Meteors can be seen on any clear night. Visible as streaks of light and usually persisting for less than a second, they occur at heights of between 75 and 100 km (47 and 62 miles). They are caused by collision with air molecules of tiny particles of interplanetary dust which enter the atmosphere at up to 70 km (43 miles) per second. The particle is heated violently and destroys itself in a blaze of incandescence.

METEOR TYPES

Sporadic meteors can appear at any time and from any direction in the sky. This photograph (*above*) shows a yellow head-on sporadic meteor seen near the Plough on 19 December, 1984. The meteor is visible as a yellowish point just to the north (above) Megrez, the top left star in the 'bowl' of the Plough.

However, the Earth frequently passes through the orbital paths of comets. At these times, particles of cometary dust which have become spread out along the orbit of the comet enter the atmosphere. These give rise to a meteor shower, with large numbers of meteors appearing over a short period.

Really bright meteors, with magnitudes of minus 5 or higher, are classed as fireballs. These are produced by particles which may be large enough to at least partially survive the fall to Earth and which may produce a meteorite fall.

METEOR RADIANTS

The particles of dust shed by the comet travel along the cometary orbit in parallel paths (see A in the diagram *below*). As a result, when particles from the same comet enter the atmosphere, they all seem to radiate from a particular point in the sky. This point is known as the 'radiant' (B). The idea of meteor radiants is clearly illustrated by looking along a straight road. The road edges, telegraph wires and so on, all of which are laid in parallel paths, seem to meet at a point on the horizon (C). This point is analogous with the radiant of a meteor shower.

There are around 20 meteor showers per year. Some are quite feeble, although around a dozen are responsible for between ten and a hundred meteors per hour at times of maximum activity which occurs when the Earth passes through the densest region of a swarm of cometary particles. Each shower is named after the point on the celestial sphere containing the radiant. For example, the Perseids, the most active shower of all, radiate from a point near Eta Persei.

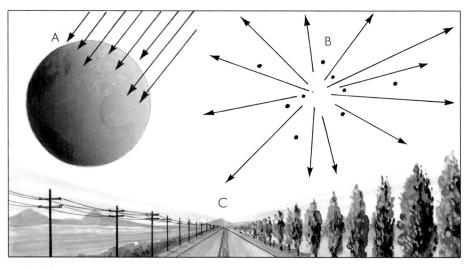

PRINCIPAL METEOR SHOWERS

SHOWER	DATE OF MAX ACTIVITY	LIMITS OF ACTIVITY	APPROX LOCATION OF RADIANT	ASSOCIATED COMET
QUADRANTIDS	3/4 JANUARY	1-6 JANUARY	10° TO S OF IOTA DRACONIS	NOT KNOWN
APRIL LYRIDS	21/22 APRIL	19-25 APRIL	BETWEEN KAPPA LYRAE AND MU HERCULIS	THATCHER
ETA AQUARIDS	5/6 MAY	24 APRIL-20 MAY	IMMEDIATELY TO N OF GAMMA AQUARII	HALLEY
PERSEIDS	11/12 AUGUST	23 JULY-20 AUGUST	3° TO NE OF ETA PERSEI	SWIFT-TUTTLE
ORIONIDS	21/22 OCTOBER	16-27 OCTOBER	BETWEEN XI ORIONIS AND GAMMA GEMINORUM	HALLEY
TAURIDS	4/5 NOVEMBER	20 OCTOBER-30 NOVEMBER	TWO RADIANTS: 10° TO W OF HYADES 12° TO NW HYADES	ENCKE
LEONIDS	17/18 NOVEMBER	15-20 NOVEMBER	5° TO NW OF GAMMA LEONIS (ALGIEBA)	TEMPEL-TUTTLE
GEMINIDS	13/14 DECEMBER	7-16 DECEMBER	IMMEDIATELY TO W OF ALPHA GEMINORUM (CASTOR)	ASTEROID 3200 PHAETHON?

N.B.: MOST OF THE NAMED STARS CAN BE IDENTIFIED ON THE MAIN SEASONAL WHOLE-SKY STAR CHARTS

METEORITES

Most particles entering the atmosphere are so small that they never reach the surface. Sometimes, however, much larger objects succumb to the Earth's gravitational pull and at least partially survive the journey through the atmosphere. Called meteorites, they appear much less frequently than meteors.

The study of meteorites has shown them to be roughly as old as the Solar System itself. Unlike meteors, many of which have their origins in comets, meteoroids (potential meteorites) travel independently through space in orbits that carry them across the Earth's path. Plotting their orbits from positional measurements taken during meteorite falls show that many of these objects originate in the asteroid belt.

Discovered in 1902 and weighing in

METEORITE FALLS

Very large meteorite falls are rare, yet when they do occur they can cause a great deal of damage, as borne out by the Arizona Meteorite Crater seen here (*below*). This huge impact crater is the result of a meteorite fall which took place thousands of years ago. Measuring over a kilometre (over ½ mile) in diameter, the floor of the Arizona Meteorite Crater lies nearly 200 m (660 ft) below the surrounding terrain. Luckily no large meteorites have fallen on densely populated areas, although fatalities resulting from smaller meteorite falls have been

at around 15 tonnes, the Willamette Meteorite, seen here (*below*) can be seen at Hayden Planetarium, New York. Iron meteorites, such as the Willamette Meteorite, are among the rarest of the three types known. Only around 5 per cent of falls are classified as irons; stony meteorites account for around 95 per cent. The rarest of all are the stony-irons.

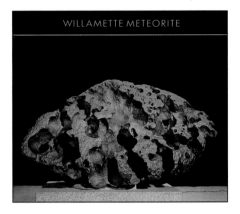

WILLAMETTE METEORITE

recorded, albeit only of animals. A meteorite fall in 1860 killed a calf in Ohio and one in 1911 finished off a dog in Egypt! Deaths to humans have been reported but not substantiated. However, in 1954 a housewife in Alabama was hit on the arm by a meteorite which came down through the roof of her house! Another near-miss account that seems to be true is that of a 19th century Swedish farmer who narrowly escaped death when a meteorite landed next to him as he was working on his land. He arranged to have the offending meteorite mounted on his tombstone where it can be seen to this day.

ARIZONA METEORITE CRATER

GLOWS IN THE SKY

Motions within the Earth's liquid-iron outer core produce a huge dynamo effect, creating electrical currents which are pushed outwards by the Earth's rotation. The result is a huge magnetic field, completely surrounding the Earth and stretching many thousands of kilometres (miles) out into space.

THE MAGNETOSPHERE

This magnetic field is known as the magnetosphere (see diagram *below*) and it is in constant reaction with energized particles from the Sun. As solar particles meet the magnetosphere, they encounter its outer boundary, called the magnetopause. A bow shock is set up as the solar wind, carrying particles which travel from the Sun at speeds of around 600 km (375 miles) per second, is suddenly slowed down.

Generally speaking, these particles are deflected around our planet by the magnetosheath which, as its name suggests, is a region surrounding the magnetosphere. Occasionally, however, some particles are pulled down through the magnetosphere, to make their way into two large, doughnut-shaped radiation belts. Discovered in 1958 by the American artificial satellite Explorer 1, they are named after James Van Allen, the principal scientist for the Explorer 1 mission. The Van Allen Belts consist of energized particles in perpetual motion between the Earth's magnetic poles. The inner Van Allen Belt is some 3,000 km (1,900 miles) thick. Containing mainly protons, it reaches a height of around 5,000 km (3,000 miles) above the Earth's surface. The outer Van Allen Belt is around 6,000 km (3,800 miles) deep and reaches a height of around 19,000 km (12,000 miles).

AURORAE

A sudden build-up of solar activity, such as that produced by large solar flares (see The Sun, pages 16-19), produces an increase in the number of charged particles leaving the Sun and, consequently, hitting the Earth's magnetosphere. The effect of this is to overload the Van Allen Belts, causing some particles to spill out into the Earth's upper atmosphere. Reactions set up between these particles and particles in the atmosphere cause nitrogen and oxygen in the atmosphere to glow, producing the displays we see as the aurorae.

Displays occurring in the northern hemisphere are known as the aurora borealis, their southern hemisphere counterparts being the aurora australis. The name aurora borealis means 'northern dawn', a term which describes very well many displays which are seen as glows on the northern (or southern) horizon. Auroral displays tend to occur near the Earth's magnetic polar regions, although they have been seen, albeit on rare occasions, at much lower latitudes. Displays generally take place at heights exceeding 100 km (60 miles) and can take on a range of colours including green, blue and red.

Displays can also take on a variety of forms with huge arcs spanning the sky, brilliant rays reaching up from the arcs and bands, formed as auroral arcs fold along their lengths. Nor is the light from aurorae constant. Changes in brightness, either slow variations, or pulsations, occurring over minutes or much more rapid fluctuations. Rapid, flaming activity produces changes which take place over just a few seconds.

This photograph of a beautiful aurora (*opposite*) is seen against a backdrop of stars in the constellation Ursa Major and was photographed from Cumbria, England in October 1981. The display resembles a huge curtain in the sky.

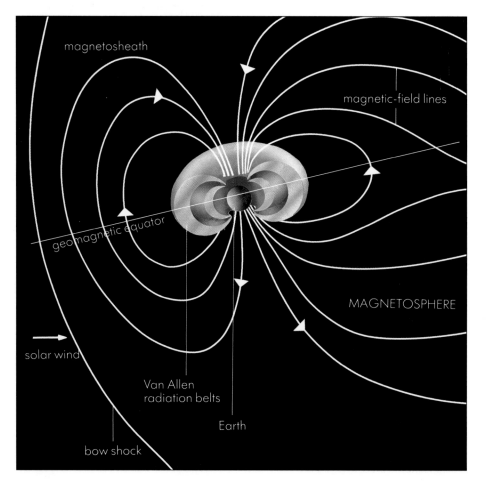

magnetosheath

magnetic-field lines

geomagnetic equator

MAGNETOSPHERE

solar wind

Van Allen
radiation belts

Earth

bow shock

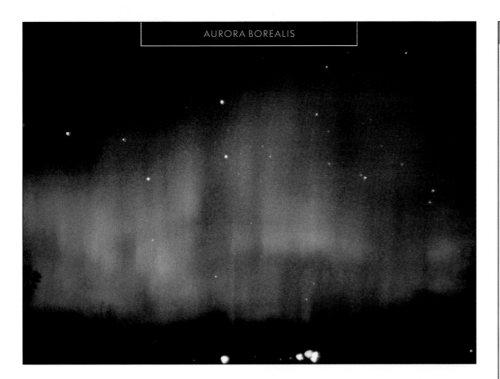

AURORA BOREALIS

THE ZODIACAL LIGHT AND ZODIACAL BAND

Unlike the aurorae, which occur within the atmosphere, the Zodiacal Light, Zodiacal Band and Gegenschein (see below) occur well beyond the confines of our planet. They are produced by sunlight reflected from dust particles scattered along the main plane of the Solar System. The Zodiacal Light takes the form of a cone of light reaching up from the horizon. The width at the base of the cone can reach 20°, its height being anything from 40° to 60° above the horizon. It is seen either in the west after sunset or in the east before sunrise. So elusive is the Zodiacal Light that photographs are rare. The picture shown here (*above right*) taken in March 1928, shows the Zodiacal Light reaching up from behind an observatory dome.

The Zodiacal Band can be regarded as an even more elusive extension of the Zodiacal Light. Its width can be anything between 5° and 10° and it can be seen, on extremely rare occasions, running along the ecliptic (the apparent path of the Sun through the sky) and extending from the top of the Zodiacal Light.

ZODIACAL LIGHT WITH STAR TRAILS

THE GEGENSCHEIN

Described as the most elusive object in the sky, the Gegenschein, or counterglow, is seen as an elliptical patch of light at the anti-solar point. This is the point on the ecliptic which is exactly opposite the Sun at the time of observation. Typically, it measures 10° × 20°, although larger displays have been seen. This photograph (*right*), taken in October 1950 through a special wide-angle camera, captures the Milky Way, which can be seen running from upper left to bottom centre of picture. Just to the upper right of centre the faint glow of the Gegenschein can also be seen.

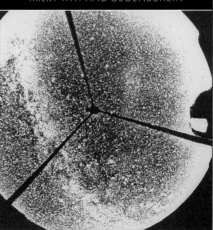

MILKY WAY AND GEGENSCHEIN

SOLAR ECLIPSES

Solar eclipses occur when the Moon passes between the Earth and Sun and the lunar shadow falls onto the Earth's surface. When viewed from the Earth, the Sun and Moon have roughly the same apparent size, in spite of the great difference in their actual diameters. As a result, during an alignment of the three bodies, the lunar disc neatly covers the Sun as seen from the Earth. The lunar shadow has two regions: a dark, central area of total shadow known as the umbra; surrounding this is an area of partial shadow called the penumbra. (Note that these have nothing to do with the umbrae and penumbrae of sunspots.)

TYPES OF SOLAR ECLIPSE

There are three types of solar eclipse: which one occurs depends upon how precisely the Earth, Moon and Sun are aligned at the time, and also upon the distance of the Moon from the Earth.

If the three bodies are exactly aligned and the lunar disc completely obscures the Sun a total solar eclipse will result (see *below top*). To an observer on the Earth's surface standing within the umbra of the lunar shadow, the Sun will temporarily disappear from view. The lunar shadow travels across the Earth's surface due to a combination of the Moon orbital motion and the axial rotation of our planet. The route taken by the umbra is referred to as the path of totality, and observers within this path will, at some stage, see a total solar eclipse (subject to cloud-cover, of course!). The total phase of the eclipse will last for up to several minutes and will be preceded and followed by partial phases.

When the Earth simply passes through the penumbra a partial solar eclipse will be seen, with only part of the solar disc being covered. A similar effect is seen by observers at either side of the path of totality during a total eclipse.

The third type of solar eclipse arises when the Moon is at or near the furthest point in its orbit from us. Because it is further away, its apparent diameter will be correspondingly reduced and it will not completely cover the Sun even during an exact alignment. The Sun will be visible as a bright ring around the lunar disc. These so-called 'annular' eclipses (*below bottom*) are named as such from the Latin word *anulus* meaning 'ring'.

EXAMPLE OF A TOTAL SOLAR ECLIPSE

It has often been said that, during total solar eclipses, day is turned into night, and indeed the sky does become quite dark with several of the brighter stars springing into view. Any of the bright planets which happen to be above the horizon at the time may also become visible. A total solar eclipse is a stunning spectacle and, in spite of the fact that the total phase is only visible from the restricted region of the path of totality, the prospect of seeing such an event prompts many astronomers to travel great distances in order to view it.

When the Sun's disc is completely covered, as seen here (*opposite left*) during the eclipse of 7 March, 1970, the glow of the solar corona can be seen surrounding the dark lunar disc. The feeble glow from the corona is usually swamped by the overpowering light from the solar disc and is visible during a total solar eclipse because the Sun's light is blotted out. Together with the corona, any sizeable prominences erupting from the solar limb may also be seen.

FACT

SUPERSTITION

As we have seen, a total solar eclipse can be an awe-inspiring spectacle, so much so that the eclipse which took place on 28 May, 585 BC was enough to bring the war between the Medes and the Lydians to a halt (see *opposite right*)! An eclipse occurred during a battle being fought in the sixth year of the war between the two nations. So shaken were the two armies by the temporary changing of day into night, that peace was hastily concluded!

TOTAL AND ANNULAR SOLAR ECLIPSES

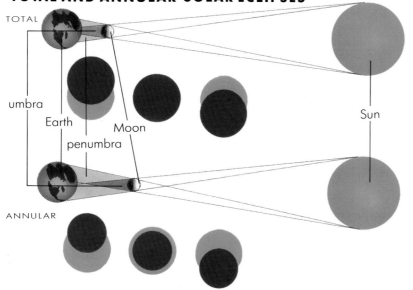

TOTAL SOLAR ECLIPSE OF 7 MARCH, 1970

SOLAR ECLIPSE OF 28 MAY, 585 BC

SOLAR ECLIPSES: 1992-2000

DATE	TYPE OF ECLIPSE	GENERAL AREA OF VISIBILITY
4/5 JAN 1992	ANNULAR	CENTRAL AMERICA, CENTRAL PACIFIC OCEAN, JAPAN
30 JUN 1992	TOTAL	SOUTH ATLANTIC OCEAN
23/24 DEC 1992	PARTIAL	JAPAN, ARCTIC
21 MAY 1993	PARTIAL	ARCTIC, GREENLAND, ICELAND, NE ASIA
13 NOV 1993	PARTIAL	ANTARCTIC
10 MAY 1994	ANNULAR	PACIFIC OCEAN, MEXICO, USA, CANADA, ATLANTIC OCEAN
3 NOV 1994	TOTAL	CENTRAL AND SOUTH AMERICA, SOUTH ATLANTIC OCEAN, SOUTH AFRICA
29 APR 1995	ANNULAR	SOUTH PACIFIC OCEAN, CENTRAL AND SOUTH AMERICA, SOUTH ATLANTIC OCEAN
24 OCT 1995	TOTAL	IRAN, INDIA, EAST INDIES, JAPAN, PACIFIC OCEAN
17 APR 1996	PARTIAL	NEW ZEALAND, SOUTH PACIFIC OCEAN
12 OCT 1996	PARTIAL	GREENLAND, ICELAND, EUROPE, NORTH AFRICA
8/9 MAR 1997	TOTAL	EAST ASIA, JAPAN
1/2 SEP 1997	PARTIAL	SOUTHERN AUSTRALIA, NEW ZEALAND, SOUTH PACIFIC OCEAN
26 FEB 1998	TOTAL	PACIFIC OCEAN, CENTRAL AMERICA, ATLANTIC OCEAN
21/22 AUG 1998	ANNULAR	INDIAN OCEAN, SOUTH AND SOUTH EAST ASIA, PACIFIC OCEAN, AUSTRALASIA
16 FEB 1999	ANNULAR	SOUTH AFRICA, INDIAN OCEAN, PACIFIC OCEAN, AUSTRALASIA
11 AUG 1999	TOTAL	NORTH ATLANTIC OCEAN, GREENLAND, ENGLAND, CENTRAL EUROPE, NORTH AFRICA

LUNAR ECLIPSES

Lunar eclipses occur at Full Moon when the Sun and the Moon are exactly opposite each other in the sky, and the Moon passes into the Earth's shadow. The Earth's shadow, like that of the Moon, consists of umbral and penumbral regions (see Solar Eclipses, pages 52-53).

TYPES OF LUNAR ECLIPSE

Total lunar eclipses (see diagram *below left*) are seen on those occasions when the Moon moves completely into the umbra, or region of darkest shadow. An observer standing on the lunar surface withing the umbra of the Earth's shadow during an eclipse of this type would see the Sun disappear completely behind the body of the Earth.

If only part of the lunar disc enters this region, a partial lunar eclipse will result (*below right*). Both these types are more readily spotted than penumbral lunar eclipses, which happen when the Moon does not enter the umbra at all, and only passes through the penumbra. From an observational point of view, it helps to know when a penumbral eclipse is due to take place as the effect of darkening on the lunar surface is only slight.

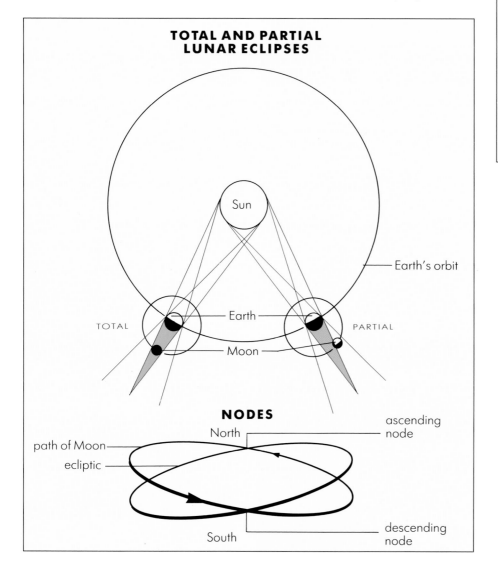

TOTAL AND PARTIAL LUNAR ECLIPSES

Sun

Earth's orbit

Earth

TOTAL

PARTIAL

Moon

NODES

path of Moon

ecliptic

North

ascending node

South

descending node

EXAMPLE OF A
TOTAL LUNAR ECLIPSE

The only reason that we see the Moon is that it reflects sunlight. During a lunar eclipse, this source of light is temporarily removed as the Moon passes into the Earth's shadow. However, even though all direct sunlight is cut off during the eclipse, the Moon can usually still be seen, although it takes on a deep coppery-red colour as sunlight is refracted onto its surface through the Earth's atmosphere. The visibility of the Moon during total lunar eclipses really depends on atmospheric conditions at the time. On rare occasions the Moon is reported to have disappeared from view altogether, as during the eclipses of 5 May, 1110 and 18 May, 1761.

Observation of lunar eclipses by the Greek astronomer Aristotle over 2,000 years ago helped him to deduce that the Earth was spherical. The Ancient Greeks knew the cause of lunar eclipses, and Aristotle noticed that each time one was seen, the shadow of the Earth on the lunar surface was curved. This could only be produced by a spherical object (the Earth) casting the shadow. The typical red coloration and the curve of Earth's shadow is clearly seen in the photograph shown here, taken during the total lunar eclipse of 18 November, 1975 (*below*).

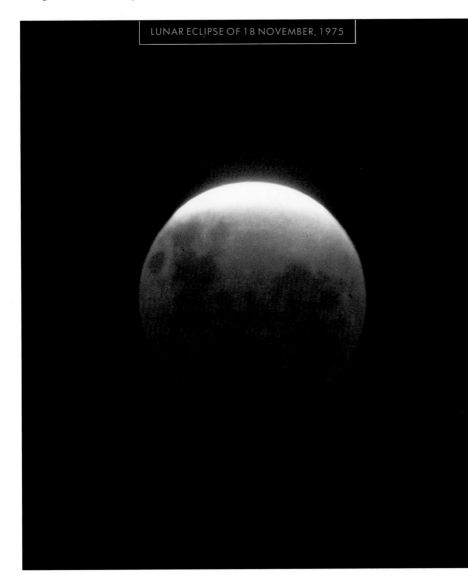

LUNAR ECLIPSE OF 18 NOVEMBER, 1975

LUNAR ECLIPSES: 1992-2000	
DATE	TYPE OF ECLIPSE
15 JUN 1992	PARTIAL
9 DEC 1992	TOTAL
4 JUN 1993	TOTAL
29 NOV 1993	TOTAL
25 MAY 1994	PARTIAL
18 NOV 1994	PENUMBRAL
15 APR 1995	PARTIAL
8 OCT 1995	PENUMBRAL
4 APR 1996	TOTAL
27 SEP 1996	TOTAL
24 MAR 1997	PARTIAL
16 SEP 1997	TOTAL
13 MAR 1998	PENUMBRAL
8 AUG 1998	PENUMBRAL
6 SEP 1998	PENUMBRAL
31 JAN 1999	PENUMBRAL
28 JUL 1999	PARTIAL
21 JAN 2000	TOTAL
16 JUL 2000	TOTAL

HOW FAR THE STARS?

THE PRINCIPLE OF TRIGONOMETRICAL PARALLAX

Distances to nearby stars are measured by trigonometrical parallax, a method which involves observing the star from two different locations and assessing its angular shift against the background of more distant stars. The principle of parallax is easy to demonstrate. Close one eye and hold up a finger at arm's length, lining it up with a suitable reference point, such as a tree. Looking at the finger through the other eye will produce an apparent movement, or shift, of your finger in relation to the tree. This is because you are viewing it from a slightly different direction. By knowing the distance between your eyes and the angle

by which your finger appears to shift, the distance to your finger can be worked out using simple trigonometry.

Astronomers apply the same principles to the measurement of stellar distances. Just as your finger appears to shift against the background when viewed alternately with each eye, a nearby star will undergo a shift against the background of more distant stars when observed from opposite points in the Earth's orbit around the Sun. By measuring the angular shift of the star and knowing the diameter of the Earth's orbit, the distance to the star can be calculated. (See diagram *right*.)

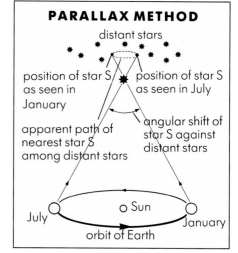

PARALLAX METHOD

distant stars

position of star S as seen in January

position of star S as seen in July

apparent path of nearest star S among distant stars

angular shift of star S against distant stars

July · Sun · January

orbit of Earth

THE PRINCIPLE IN PRACTICE

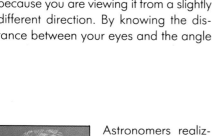

Astronomers realized that trigonometrical parallax could be used to assess stellar distances long before the method was first used successfully, although early attempts were thwarted either through lack of observational skills on the part of the observer or because of inferior equipment. Stars were once thought to be fixed in space, although observations carried out by the English astronomer Edmund Halley revealed that a number of bright stars had changed position over the centuries since star catalogues were first drawn up. It seemed logical to assume that the closer a star was to Earth, the greater would be its angular shift across the sky. The shift in a star's position is a result of the actual movement ('proper motion') of that star through space.

The German astronomer Friedrich Wilhelm Bessel, pictured here (*inset*), noticed that the star 61 Cygni (see accompanying chart *right*) had a substantial proper motion, amounting to an

angular shift against the background stars of 5·2 seconds of arc per year. In 1838 he selected 61 Cygni as the star whose distance he would attempt to measure using trigonometrical parallax. Bessel had access to good quality telescopes and equipment, which allowed him to measure 61 Cygni's tiny parallax shift; this amounted to around 0·29

seconds of arc. Application of trigonometry enabled him to work out that 61 Cygni was located at a distance of 10·3 light years. Comparison with the currently accepted value of 11·1 light years is testament to Bessel's skills as an observer, his efforts finally giving astronomers an idea of the distribution of stars in the solar neighbourhood.

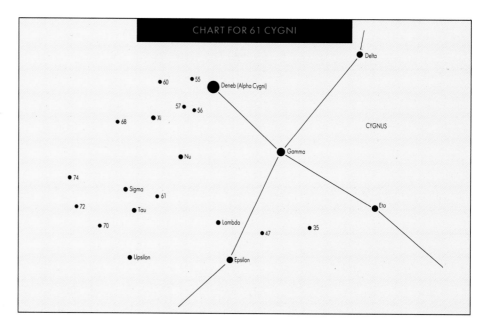

CHART FOR 61 CYGNI

THE COMPARISON METHOD OF CALCULATING STELLAR DISTANCES

The one drawback with applying trigonometrical parallax to the measurement of stellar distances is that it is only effective for stars within a distance of around 70 light years. Beyond this, the angular shifts become too tiny to be measured accurately. In order for more distant stars to have their distances measured, astronomers draw up comparisons between stars of similar types, one of which is at a known distance and the other further away.

This method of comparison assumes that stars of similar types will have similar actual luminosities and allows us to assess the distance to the more distant star by comparing its apparent and actual brightnesses. It is rather like calculating how far away a distant searchlight is by knowing the true brightness, or power, of the light.

Calculation of stellar distances reveals that the constellations are generally made up of stars that lie at different distances from us, the patterns we see being due to nothing more than chance alignments of the stars. The constellation Corvus is a good example (see diagram *below*). The distances to its principal stars vary from just 68 light years in the case of Alpha Corvi to 290 light years for Beta. The stars that form Corvus would be differently arranged if viewed from other locations in space.

DISTANCES OF STARS IN CORVUS

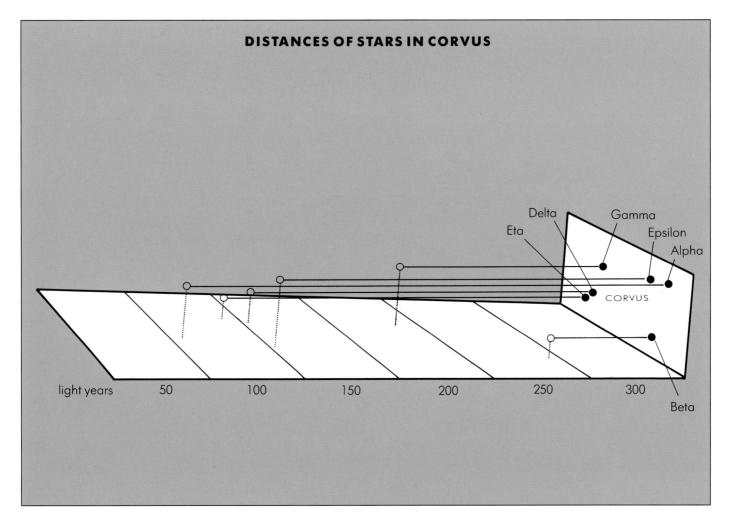

THE MESSAGE OF STARLIGHT

MAGNITUDE SYSTEMS

In around 150 BC the Greek astronomer Hipparchus divided the stars up into six classes of apparent brightness, the brightest stars being ranked as first class and the faintest as sixth. In 1856 Norman Pogson refined Hipparchus's system by classing a 1st magnitude star as being 100 times as bright as one of 6th magnitude, giving a difference between successive magnitudes of $\sqrt[5]{100}$ or 2·512. A star of magnitude 1·00 is 2·512 times as bright as one of magnitude 2·00, and 6·31 (2·512 × 2·512) times as bright as a star of magnitude 3·00 and so on. The same basic system is used today, although modern telescopes enable us to determine values to within 0·01 of a magnitude. Negative values are used for the brightest objects including the Sun (−26·8), Venus (−4·4 at its brightest) and Sirius (−1·42). This system classifies stars and other celestial objects according to how bright they appear to the observer and offers no real guide as to their true luminosity. Clearly a system is required that provides an accurate guide to the actual brightness of stars.

Astronomers have devised such a system. Defining the absolute magnitude of stars expresses the apparent magnitude a star would have if placed at a standard distance of 10 parsecs (32·6 light years). A parsec (3·26 light years) is the distance at which a star would subtend an angle of parallax of 1 second of arc (see How Far The Stars, pages 56-57). The absolute magnitude of a star is a far more accurate assessment of its true luminosity. For example, if the star Sirius (apparent magnitude −1·42) were placed at a distance of 10 parsecs, its magnitude would be +1·4. Compare this with Deneb (Alpha Cygni). Deneb appears to us as a star of magnitude 1·26, nearly three magnitudes fainter than Sirius. However, Deneb shines from a distance of 1,600 light years, as compared with Sirius which lies 8·8 light years away. The absolute magnitude of Deneb is −7·5, showing its true luminosity to be thousands of times that of Sirius.

THE ELECTROMAGNETIC SPECTRUM

Stars and other celestial objects emit electromagnetic radiation. This is a waveform and ranges from the very short wavelength gamma rays, X-rays and ultra-violet, through visible light and on to infra-red, microwave and radio emissions (see *right*). The distance between successive wave crests is referred to as the wavelength, those of gamma, X-ray and ultra-violet radiation and visible light generally being expressed in units of length called the Ångstrom. This is abbreviated Å and named after the Swedish physicist Anders Jonas Ångstrom. 1 Å is equal to 10^{-10} metres. Other units of length used are the micron (μ), 1μ being equal to 10^{-6} metres. Wavelengths longer than 10 mm are expressed in centimetres, metres or kilometres, the latter used for very long wavelength radio waves.

Not all wavelengths are able to penetrate the Earth's atmosphere. For example, gamma rays, X-rays and most ultra-violet radiation is effectively blocked by the atmosphere. In order to carry out

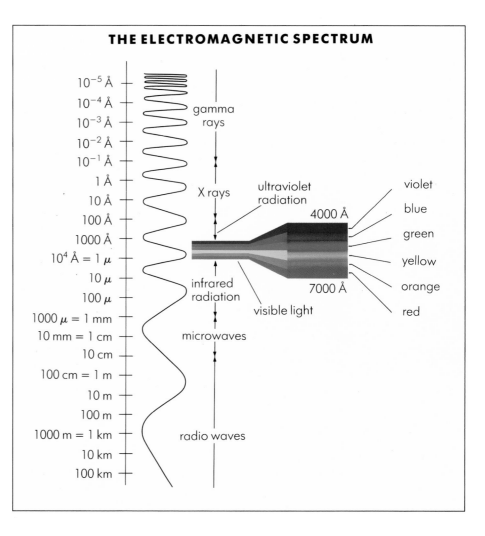

THE ELECTROMAGNETIC SPECTRUM

SPECTRA OF VEGA (ALPHA LYRAE)

The electromagnetic spectrum is made up of the complete range of electromagnetic radiation of which visible light forms only a tiny part. Passing the visible light received from a star through a prism will split it up into its constituent colours which range from short-wavelength violet to long-wavelength red. The multicoloured band of light produced by this process is called a spectrum.

Examination of the spectrum of a star will show that it is made up of two separate spectra. Several exposures showing the light from Vega (Alpha Lyrae) broken into its constituent colours are shown here (below). The high density gases making up a star produce a continuous spectrum, an unbroken sequence of colours ranging from violet to red. However, the light reaching us from the star comes not only from the star itself but also from the low-density gases surrounding it; in other words, from the stellar atmosphere. Gas of low density also produces a spectrum, but of a different

THE CLASSIFICATION OF STELLAR SPECTRA

In 1863 the Italian astronomer Angelo Secchi published the first classification of stellar spectra. He divided stellar spectra into five groups based both on spectral characteristics and colours of the stars. Secchi's classification was improved upon half a century later. Known as the Harvard Classification, this divided spectra into a greater number of classes, each denoted by a letter from the sequence O, B, A, F, G, K, M, R, N and S.

The Harvard system classifies stars according to their temperatures. It ranges from the hottest O stars, with surface temperatures in excess of 35,000°C (63,000°F) down to the coolest M, R, N and S stars, with surface temperatures of only 3,000°C (5,400°F) or less. These broad classifications are further subdivided by the insertion of a number, ranging from 1 to 9, after the letter. Examples include A2 (Deneb), K5 (Aldebaran) and G2 (the Sun).

THE HERTZSPRUNG-RUSSELL DIAGRAM

The Hertzsprung-Russell (HR) diagram (see *below*) classifies stars according to their spectral types or temperature and their absolute magnitudes or luminosity. Bright stars are plotted to the top of the HR Diagram, fainter stars to the bottom. It is clear that the colour of a star is a good indicator of its surface temperature, the hottest blue-white or white stars of spectral classes O, B and A being situated at the left side, the cooler red stars of classes K and M appearing at the right.

The Sun lies on the Main Sequence, a band of stars running from upper left to lower right. Main Sequence stars are currently in the hydrogen-burning phase of their evolution (see Stellar Evolution, pages 60-63). Groups other than the Main Sequence are also evident, including the huge, cool supergiant stars. Although less light is emitted per unit of surface area, their surface areas are extremely large. Supergiant stars, therefore, are very bright. Of course some supergiants, such as Rigel in Orion and Deneb in Cygnus, have high surface temperatures, their positions near the upper left-hand corner of the diagram reflecting this. White dwarfs, on the other hand, have high surface temperatures but are fairly faint, a relationship indicative of their small diameters.

SPECTRA OF VEGA

type. This will appear as a series of individual lines, each line being the result of the effects of a particular element. This so-called emission spectrum can be seen superimposed upon the continuous spectrum as revealed here. The emission lines appear dark in comparison to the much brighter continuous spectrum.

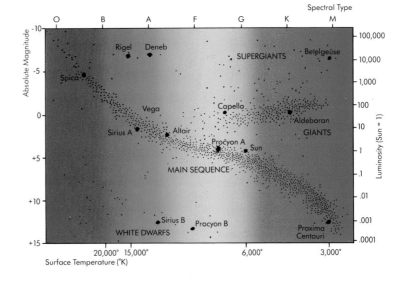

THE HERTZSPRUNG-RUSSELL DIAGRAM

STELLAR EVOLUTION 1

The evolution of stars from their formation through to eventual demise takes place over huge time spans. The Sun has been shining as at present for some 4,500 million years and will continue to do so for a similar length of time before any significant changes take place within it. Faced with these huge time spans it may at first seem an impossible task to piece together the life cycle of a star.

Yet astronomers, by carefully observing many different stars, have managed to link the different events in a star's life that take it from a collapsing clump of nebulosity through to a white dwarf, neutron star or black hole.

THE BIRTH OF STARS

Star formation takes place inside huge clouds of gas and dust scattered throughout the spiral arms of our and other galaxies. The spiral galaxy NGC 300, pictured here (*below*), is a member of the Sculptor Group of Galaxies. Knots of nebulosity can be seen in the disc of this galaxy, these regions being the scenes of star formation. In order for star formation to commence, the particles of gas within the nebulae must collapse to form clumps of matter. This collapse, known as fragmentation, takes place within denser portions of the nebula and is instigated by any of a number of different processes.

In some cases the shock waves from a nearby supernova explosion may disrupt the gas within the spiral arms. Other possibilities include the effects of radiation from nearby young, hot stars and the rotation of the galaxy itself. As a galaxy rotates and the spiral arms sweep around, the impact between nebulae and other interstellar matter produces localized condensations.

YOUNG STARS

NGC 7538, located some 7,000 light years away in the constellation Cepheus, is a region of massive star formation within our own Galaxy. This infrared image (*below*) shows a reddish object, visible towards the lower left of picture. This is a young star which is still enveloped in the cloud from which it formed. Light from this star is being scattered by dust grains within the cloud, this scattering being visible as a bluish extension to its lower right. The huge cloud at top right of picture is a nebula in which the hydrogen is being ionized by strong ultraviolet radiation from a number of young, massive stars embedded within it. Emission nebulae of this type normally appear red at optical wavelengths (see Nebulae, pages 84-85), this infrared photograph showing it as a blue patch.

As fragmentation continues, the densities of the clumps increase as material slowly collects together, this in turn producing increased temperatures at their cores. The heat generated is unable to escape due to the build-up of density, leading to a growth in outward-acting thermal pressure. This eventually reaches a point whereby it balances the inward pull of gravity. The process of fragmentation comes to a halt, the nebula now consisting of a number of balanced regions astronomers call protostars.

SPIRAL GALAXY NGC 300

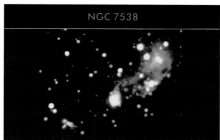

NGC 7538

STABLE STARS

The path taken by the protostar from here depends upon its mass. Those of around one solar mass start to undergo hydrogen burning whereby hydrogen is converted to helium. These reactions are the same as those taking place at the heart of the Sun (see The Sun, pages 16-19), and are triggered off once their core temperatures reach 15 million °C (27 million °F). Heat energy acts outwards, preventing further collapse. The star is now stable and proceeds to shine through the release of energy.

The star now enters the longest phase of its life and moves onto the Main Sequence of the Hertzsprung-Russell Diagram (see The Message of Starlight, pages 58-59). A star comparable in mass to the Sun will produce energy for around 10,000 million years before proceeding to the next stage of its evolution. Much more massive stars require greater amounts of thermal energy to prevent their correspondingly greater inward collapse. In order to create the huge amounts of heat required they burn their fuel at a much faster rate, thereby reaching the next stage of development far earlier. Stars of lower masses than the Sun burn their fuel much more economically and remain on the Main Sequence for much longer periods.

DECLINING STARS

As the phase of hydrogen burning draws to an end, the star enters the next stage of its evolution.

At this point, the star will have a helium-rich core. Most of the energy produced by the star at this stage originates in a shell surrounding the core. Hydrogen burning continues in this shell and these continuing reactions ensure that the star remains on the Main Sequence for a few more millions of years. Eventually the shell is the only place where hydrogen burning takes place; the core is depleted of hydrogen and ceases to produce thermal energy. At this stage, the star collapses and as a result the core heats up, thereby balancing the inward collapse.

RED GIANTS

As the core becomes compressed and its temperature increases, more energy is produced. This energy has the effect of pushing the star's outer layers into space, swelling it to many times its original diameter. The star becomes a red giant. This computer-enhanced image (*below*) shows a halo of gas around the red giant star Betelgeuse (Alpha Orionis),

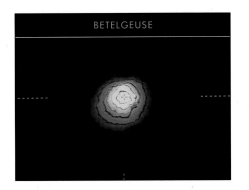

BETELGEUSE

the star marking the right shoulder of Orion. Following the expansion of the star, its outer layers cool to temperatures of around 3,000°C to 4,000°C (5,400°F to 7,200°F), the star glowing red as a result. Our Sun will undergo this sequence of events in around 5,000 million years. Although its surface temperature will be greatly reduced, the Sun's diameter, and consequently its surface area, will be considerably larger than it is now, producing roughly a hundred-fold increase in its luminosity.

Although the outer layers of a red giant are very tenuous, compression at the core increases as helium falls on to the core from the reactions taking place in the hydrogen-burning shell. Eventually the temperature reaches the critical value of 100 million °C (180 million °F), triggering further reactions. Helium is converted into carbon; eventually the helium fuel runs out. Low mass stars have too little mass to produce sufficient gravitational pressure to instigate further reactions. They collapse and cool, ending their careers as white dwarfs, unlike the more massive stars which undergo further collapse and additional nuclear reactions, thereby continuing to shine.

PLANETARY NEBULA

Visible as shells of gas surrounding hot central stars, planetary nebulae are formed as the outer layers of a star are thrown off into surrounding space prior to the star itself commencing the collapse into a white dwarf. These expanding shells of gas eventually dissipate into the interstellar medium. The average lifespan of a planetary nebula is only several tens of thousands of years. Planetary nebulae derive their name from their disc-like appearance in small telescopes, the description first being coined by William Herschel in 1785. Among the hundreds of planetary nebulae known are many fine examples including NGC 3132 in Vela seen here (*below*).

Following the ejection of its outer layers, the inner regions of the star are exposed. The surface temperatures of these stars can be anything up to 100,000°C (180,000°F) or more. The copious amounts of ultraviolet energy released by the star ionize the gas in the surrounding nebula causing it to shine at visible wavelengths.

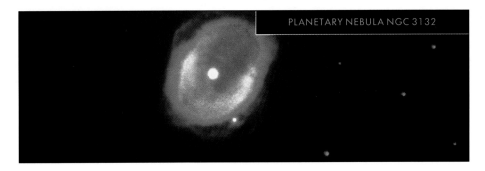

PLANETARY NEBULA NGC 3132

WHITE DWARFS

Once nuclear reactions cease, there is no outward thermal pressure. Gravity takes over and the star collapses to form a white dwarf. Stars of less than 1·4 solar masses collapse to form objects with diameters of only a few thousand kilometres (miles). (Stars of greater initial mass can undergo the same process provided they lose some of their material, perhaps through the formation of a planetary nebula.)

The result of the collapse is that their component atoms become tightly packed and the star attains an incredible density. The material in these stars ends up at something approaching a million times the density of water. The only energy produced now is that created during the collapse. The white dwarf continues to shine although the energy supply is eventually completely depleted. The once-bright solar mass star ends its career as a dead star or black dwarf.

White dwarf stars are fairly common. One of the most famous white dwarf stars can be located in Canis Minor (see accompanying chart *right*). Procyon B, the companion star of the bright star Procyon (Alpha Canis Minoris), is a remarkable star with a luminosity only 1/15,000 that of Procyon, although its mass is some 60-70 per cent that of our Sun and its diameter is only a little more than twice that of the Earth.

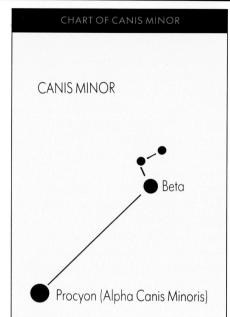

CANIS MINOR

Beta

Procyon (Alpha Canis Minoris)

SUPERNOVAS

Stars with high masses end their careers in a much more awe-inspiring way than low mass stars, the huge amounts of material within them ensuring that a prolonged series of nuclear reactions takes place. The correspondingly high amount of gravitational compression that results produces an extremely hot core. The temperatures attained allow heavier and heavier elements to undergo nuclear fusion. In stars of 25 solar masses, the internal temperature can reach 3,000 million °C (5,400 million °F). Each round of nuclear reactions involves the fusion of the remnants of the previous reaction into a heavier element.

The chain of events will cease when silicon burning forms an iron core, the iron produced being unable to fuel further reactions. At this stage, the core itself is surrounded by shells in which different elements are burning, the cross-sectional appearance of the star at this stage resembling that of an onion! Fresh iron is continually being deposited upon the core as a result of the silicon burning going on just above it. Eventually, the mass of the core becomes so great that it begins to collapse. The temperature at the core climbs alarmingly over a very short period of time. Within less than a second, the pressure becomes so great

that the collapse halts. Meanwhile, the outer layers of the star, no longer supported by nuclear reactions, fall down onto the core at colossal speeds. As they impact the core, enormous temperatures and pressures are produced which cause the material to literally bounce back, creating a wave of material that travels out towards the surface of the star. Because of the decreasing resistance experienced by the wave, its speed increases. It eventually reaches the surface, and the outer layers of the star are thrown out into space, producing the brilliant event we know as a supernova. For a time the brightness of the exploding star can rival the total light output of the galaxy in which it occurs!

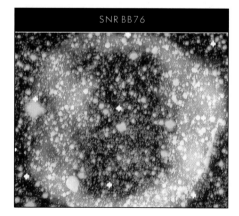

SNR BB76

There are two types of supernova, each of which originates in a different way. Type II supernovae arise from the collapse of very massive stars and are described above. Type I supernovae result from binary systems comprised of a red giant and a carbon-rich white dwarf. Material is dragged from the red giant onto the white dwarf, resulting in a gaseous layer being built up on its surface. This build-up of material eventually takes the mass of the white dwarf above 1·4 solar masses. The carbon ignites and the star is destroyed in a very short time. Type II supernovae are powered by gravitational energy, whereas nuclear energy powers Type I supernovae, the peak brightness of which generally exceeds that of Type II.

Supernova explosions have been widely observed in other galaxies, although few have been witnessed within the Milky Way. One of the most famous of these was the supernova which appeared in 1572. This huge explosion, observed and recorded by the Danish astronomer Tycho Brahe, resulted in the dying star throwing off most of its material into space. The outwardly-expanding remnant of this supernova explosion, visible today and shown here (*left*), is known as SNR BB76.

NEUTRON STARS
AND BLACK HOLES

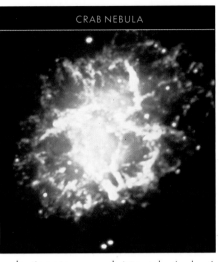

CRAB NEBULA

Depending upon its mass, a supernova remnant may collapse into either a neutron star or black hole. The core of the star which gave rise to the Crab Nebula, shown here (*right*), collapsed to form a neutron star. The Crab Nebula was the result of probably the most famous supernova, which appeared in the constellation Taurus in 1054. Although the explosion took place around 6,000 light years away, the event was easily visible to Earth-based observers, a testament to the awesome energy released by a supernova. During these events, anything up to 10 solar masses of material may be jettisoned into surrounding space.

Stars of between 1·4 and 3 solar masses suffer such high gravitational collapse that the protons and electrons within the star are crushed together to form neutrons. A neutron star, typically only several tens of kilometres in diameter, is the result. These objects have densities approaching 900,000 tonnes/cm^3 (15 million tonnes/in^3)!

The existence of neutron stars had been suggested several years before the discovery of the first example in 1967. An object emitting extremely regular radio pulses was detected at radio wavelengths, the interval between each being measured to a precise 1·3370109 seconds. The object, thought to be a pulsating star, was christened a 'pulsar'. Many more pulsars have since been found, the most famous of which lies at the heart of the Crab Nebula. This star is now known to be emitting radio pulses synchronous with its rapid axial rotation (equal to 30 times per second). It is unusual in that it is one of the very few neutron stars which have been identified at optical wavelengths.

Even more exotic are the remnants of stars greater than eight solar masses. The gravitational collapse of such stars is complete, producing a sphere of ultimate density. So great is the gravitational pull of the resulting object that its escape velocity exceeds the speed of light. Because light is unable to escape we are unable to actually see the remnants of the most massive stars. The object formed is called a collapsar (short for 'collapsed star'). Its incredible gravitational influence weakens with increasing distance until a point is reached where light can escape. This is the event horizon, which encloses the black hole, a zone permanently hidden from view.

In spite of their invisibility, astronomers think that they have detected black holes through the study of certain binary star systems. By observing the orbital motions of binary systems, we are able to calculate the masses of the member stars. One such binary, Cygnus X-1, is a powerful emitter of X-rays, radiation that would be emitted in the event that material was being pulled from one star onto the surface of a much denser companion. As the material collected is pulled down it gets very hot. Hot gas is a source of X-rays, and many X-ray sources have been identified as binary star systems with a neutron star component. However, study of the Cygnus X-1 system (one of the strongest X-ray sources in the sky) has shown that the companion of the visible supergiant component is much too great for it to be a neutron star. Cygnus X-1 is a prime candidate for playing host to a black hole, a conclusion that must await developments in observational techniques for verification.

FROM SUPERNOVA
TO STAR BIRTH

The heavy elements ejected into space by supernovae mingle with the interstellar medium. This photograph (*right*) shows the north central portion of the Veil Nebula in Cygnus, a huge loop of matter ejected from a supernova explosion thought to have taken place around 30,000 years ago and now dissipating into space. As with other supernovae remnants, some of the material from the Veil Nebula will eventually find its way into interstellar clouds that will in turn collapse to form new stars.

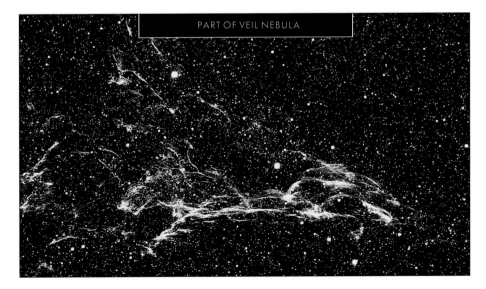

PART OF VEIL NEBULA

CONTENTS

HOW TO USE THE STAR CHARTS

The second half of the book presents a guide to selected areas of the sky and a number of deep-sky objects within the reach of the naked eye, binoculars or a small telescope. Examples of each class of deep-sky object (each class is also described in detail in this part of the book) are given, and a selection of charts are included for both northern and southern hemisphere observers.

STAR-HOPPING

Location of many of the objects described on the following pages is made by the star-hopping method, a means of deep-sky object location becoming increasingly popular among amateur astronomers. The idea behind star-hopping is simple and straightforward, and involves the location of objects by following lines or patterns of stars from 'guide stars' which are used as starting points. The guide stars are relatively bright and easily identified from either the wide-field charts accompanying many of the finder charts,

such as this one depicting part of Virgo, or from the main whole-sky charts (see Contents, page 65).

To locate your deep-sky target – in this case the Sombrero Galaxy (M104) – first of all identify the guide star or stars. In this case, we have a trio – Psi, Chi and 21 Virginis, all three of which are easily spotted by moving west from the bright star Spica (Alpha Virginis). Bring one of the guide stars into the field of view of your binoculars or, when carrying out a telescopic search, line it up with the

cross-hairs in the finder. By 'hopping' from one star to another, such as from Psi, through Chi and 21 Virginis in the direction indicated by the broken line on the accompanying M104 finder chart, you can follow a path to the object being sought. In the case of M104, this galaxy lies just off a line between 21 Virginis and Sigma 1669.

Once the object has been identified, you can draw lines on the finder charts to show the route taken to the object in question. This will make the job of location

SOMBRERO GALAXY

easier next time around. Getting used to seeking out deep-sky objects in this way can only increase your knowledge of the night sky, each repeated observation of a particular object gradually building up your knowledge of, and aquaintance with, the region of sky around it and the patterns of stars that it contains.

As with the vast majority of the objects described in the following pages, a photograph of the object covered appears here. The image of M104, shown here, was obtained at a professional observatory; a number of the charts are accompanied by photographs obtained with amateur equipment, and show the object more or less as it would appear through moderate optical aid.

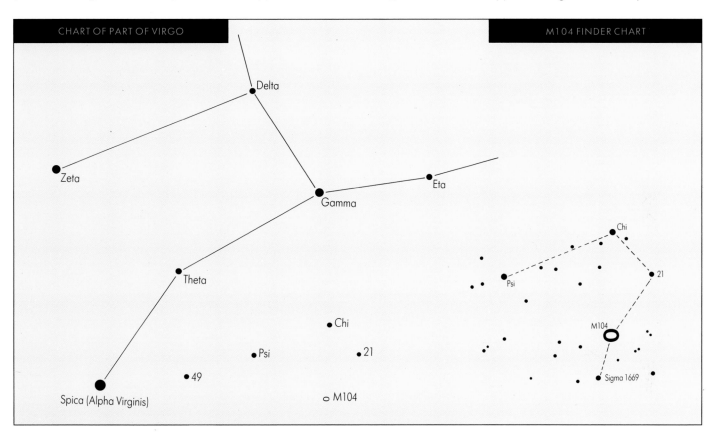

CHART OF PART OF VIRGO

M104 FINDER CHART

MILKY WAY IN CARINA AND MUSCA

NAKED-EYE OBSERVATION

Although deep sky objects are generally observable only with optical aid, a number are bright enough to be visible with the unaided eye. Notable examples include the Hyades and Pleiades open star clusters in Taurus (pages 82-83) and the Orion Nebula (pages 80-81). This wide field view of the Milky Way in Carina and Musca contains several interesting naked-eye objects including the Coalsack nebula, a huge cloud of absorbing matter located near the left margin. The Eta Carinae Nebula, the most luminous region of ionized hydrogen known in the Milky Way, is prominent to the right of centre while many open star clusters can be seen scattered across the picture.

The conspicuous shape of the Plough dominates this chart of the sky close to the north celestial pole. Merak and Dubhe, the two stars marking the end of the 'bowl' of the Plough point the way to Polaris, the Pole Star, in nearby Ursa Minor (the Little Bear). Continuing the line from Merak and Dubhe through the Pole Star and roughly as far again leads our gaze to the prominent shape of Cassiopeia, a group depicting Queen Cassiopeia of Ethiopia. Other constellations visible here include that depicting Cepheus, husband to Cassiopeia, and Draco (the Dragon), the latter winding its way around the north celestial pole.

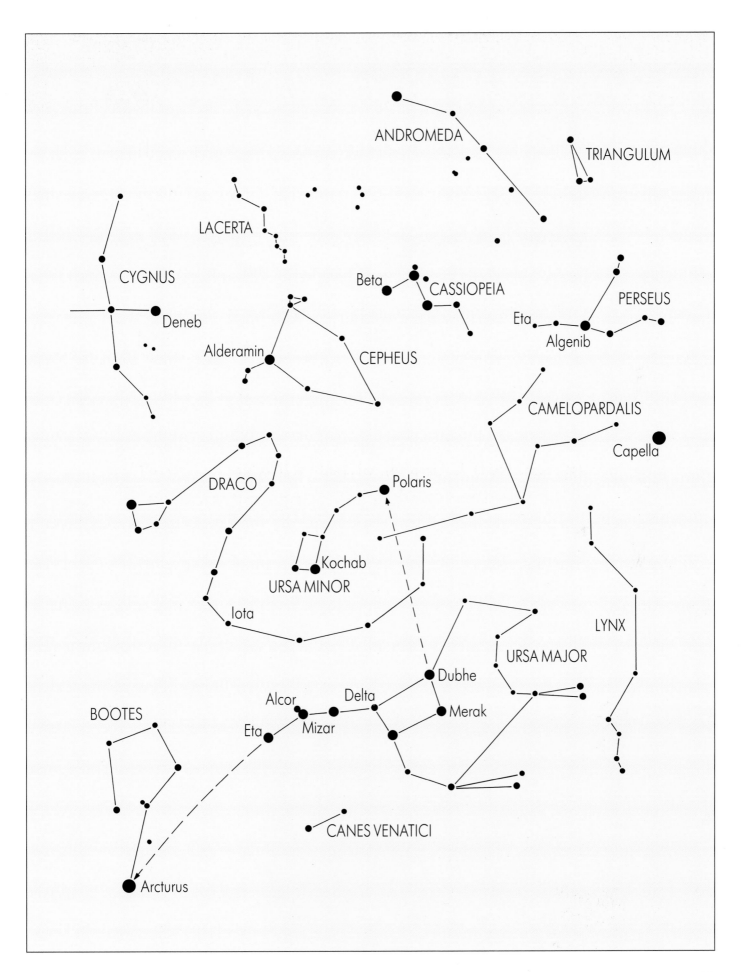

URSA MAJOR

Ursa Major (the Great Bear) is a large, sprawling constellation, although most of its stars are rather faint and difficult to pick out unless the sky is really dark and clear. The familiar pattern of stars we call the Plough is really only a part of Ursa Major, the rest of the group spreading out to the south and west. The accompanying chart (*below left*) shows the Plough together with many of the other stars in the Great Bear. The names of the stars in the Plough are given here as they will be referred to elsewhere on this page.

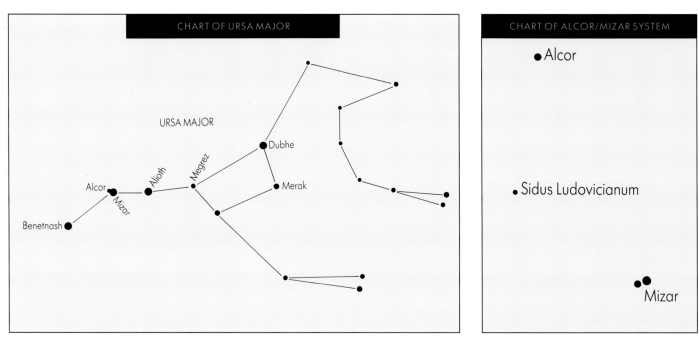

CHART OF URSA MAJOR

URSA MAJOR

Dubhe
Megrez
Alioth
Merak
Alcor
Mizar
Benetnash

CHART OF ALCOR/MIZAR SYSTEM

Alcor

Sidus Ludovicianum

Mizar

THE PLOUGH

The Plough (see *below*) can be seen almost overhead during spring evenings from mid-northern latitudes, moving to a position high over the north-western horizon by summer. Autumn evenings find the group low in the northern sky while by winter the group is visible in the north-east, making its way towards the overhead point, which it will reach by the following spring.

PLOUGH

ALCOR AND MIZAR

A close look at the middle star in the 'handle' of the Plough will show that it is in fact a double, formed from the two stars Alcor (magnitude 4·01) and Mizar (magnitude 2·09). Those with good eyesight will be able to resolve both stars with the naked eye under clear skies. A small telescope will reveal a third star forming a triangle with Alcor and Mizar (see chart *above*). This 8th magnitude object is known as Sidus Ludovicianum, a name given to it by the eccentric German astronomer J G Liebknecht some two hundred years ago. He thought he had discovered a new planet and named it after his sovereign, the Landgrave Ludwig of Hessen-Darmstadt. Liebknecht's 'discovery' was erroneous although the name he gave it has stuck! Telescopes will reveal that Mizar itself is a close double, with magnitude 2·3 and 4·0 components lying 14·4" apart.

POINTERS TO THE NORTHERN SKY

The Plough is a marvellous pointer to many other groups in the northern sky, notably as an aid to finding Polaris, the Pole Star. Polaris is the brightest member of Ursa Minor (the Little Bear), and is found by extending a line from Merak, through Dubhe and on until you reach the first reasonably bright star. The rest of Ursa Minor, resembling a smaller, inverted version of the Plough, can easily be seen in the accompanying photo-graph stretching away from Polaris. Polaris lies very close to a point in the sky called the North Celestial Pole. The Earth's axis is aligned with this point (and a corres-ponding point in the southern sky known as the South Celestial Pole – see pages 118-119). All the stars in the sky appear to rotate around the celestial poles, an effect caused by the Earth's axial rotation and the resulting apparent motion of the stars across the sky.

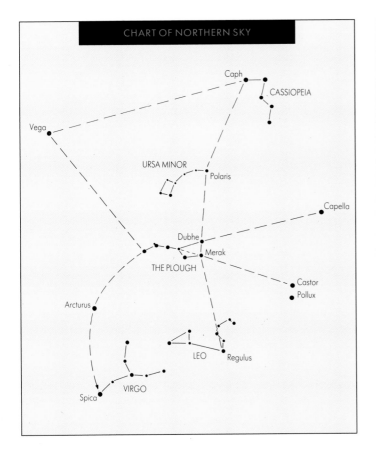

CHART OF NORTHERN SKY

URSA MINOR

Extending the line through Polaris and about as far again will bring you to Cassiopeia, a prominent W-shaped group of stars lying on the opposite side of Polaris to the Plough. The relative positions of the Plough and Cassiopeia mean that when the Plough is high in the sky, Cassiopeia is found low over the northern horizon, and vice versa. They perform annual cartwheels around the North Celestial Pole.

Another pair of 'opposites' are the two brilliant stars Vega (Alpha Lyrae) and Capella (Alpha Aurigae). Capella is located by extending a line from Megrez through Dubhe, while Vega forms a large triangle with Caph (Beta Cassiopeiae) and Benetnash. When above the horizon, these stars are un-mistakable and will be found without dif-ficulty. Capella is located at or near the overhead point during winter evenings, Vega occupying this position during the summer months. Auriga is covered in more detail on pages 78–79 and Vega on pages 98-99.

The bright pair of stars Castor and Pollux, the leading members of the con-stellation Gemini (the Twins) are located by extending a line from Megrez through Merak. Gemini is described more fully on page 83, and is best placed for observation from the northern hemis-phere during late-winter/early spring when it will be high in the south during the evenings. During the following few months Leo (the Lion) and Virgo (the Virgin) ride high over the southern hor-izon. Leo is found by extending a line from Polaris through Dubhe and Merak and down towards the south as shown. Virgo can be picked out by following the curve of the Plough 'handle' down past Arcturus (Alpha Bootis) and on to Spica (Alpha Virginis).

DRACO

Although not appearing on the general chart, the region around the North Celestial Pole contains several fainter groups including Cepheus (see page 73) and Draco (the Dragon). The Head of Draco lies a little to the north of Vega, as shown in the accompanying chart (*below*), from where it winds its way around the North Celestial Pole to a point near Polaris. The photograph (*inset*) includes the two stars Kochab and Pherkad (Beta and Gamma Ursa Minoris), seen towards the upper right of picture with Thuban (Alpha Draconis) located near the right hand margin. From here much of the group can be traced out by comparing the photograph and chart.

Draco contains several double stars within the reach of small telescopes, including the wide pair Kuma (Nu Draconis) which is also resolvable in binoculars. Here we have what is widely regarded as one of the most striking doubles in the sky, its magnitude 5·0 and 5·0 white components separated by 61·9″.

Psi Draconis has magnitude 5·0 and 6·0 components located 30·3″ apart and the pair can easily be resolved through a small telescope. Slightly more difficult is 17 Draconis. This is formed from 5·0 and 6·0 stars situated just 3·2″ from each other. In the same field as 17 is 16 Draconis, a 5th-magnitude star located 90″ away and forming a nice triple object. Another double worth seeking out is the double star Omicron Draconis, its orange and blue magnitude 4·7 and magnitude 7·5 components lying 34·2″ apart, well within the reach of a 75-mm (3-in) telescope.

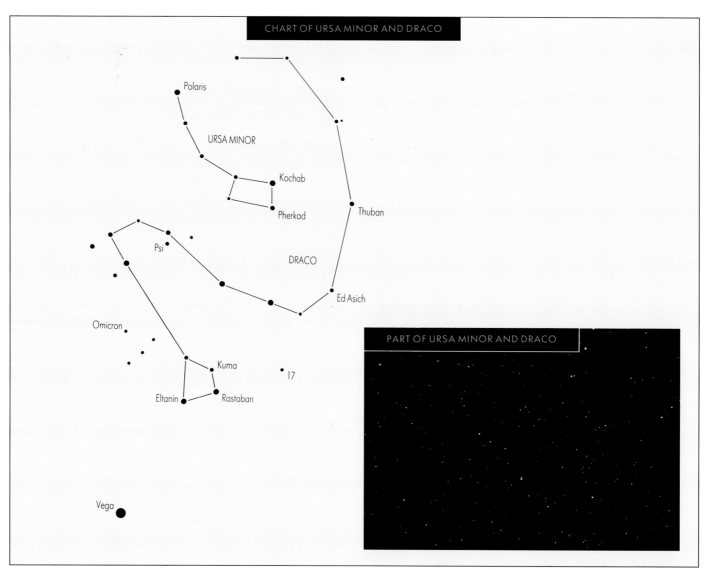

CHART OF URSA MINOR AND DRACO

Polaris

URSA MINOR

Kochab

Pherkad

Thuban

Psi

DRACO

Ed Asich

Omicron

Kuma

17

Eltanin   Rastaban

Vega

PART OF URSA MINOR AND DRACO

CASSIOPEIA, CEPHEUS AND LACERTA

This photograph (*below left*) shows the 'W' of Cassiopeia at bottom together with most of the neighbouring constellations Cepheus (top left) and Lacerta (seen faintly near upper right). The area covered by the photograph is highlighted on the accompanying chart (*below right*) which should help you pick out the regions of Cepheus and Lacerta not included on the photograph.

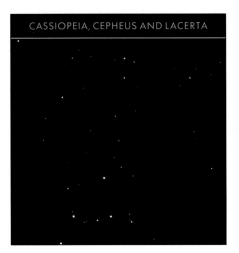

CASSIOPEIA, CEPHEUS AND LACERTA

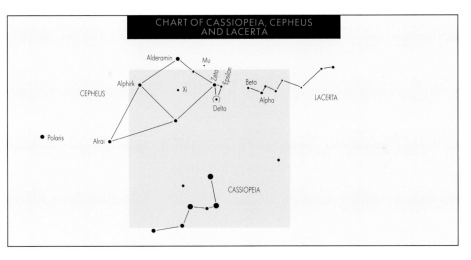

CHART OF CASSIOPEIA, CEPHEUS AND LACERTA

DOUBLE AND VARIABLE STARS IN CEPHEUS

The double star Xi Cephei is located at a distance of around 80 light years and found at the centre of the main quadri-lateral of stars forming this group. The magnitude 4·4 and 6·5 components are separated by 7·7″, corresponding to an actual separation of nearly 200 times that between the Earth and Sun. Both members are resolvable in telescopes of 75-mm (3-in) aperture or more.

The variable star Delta Cephei, which forms a tiny triangle with Zeta and Epsilon, is the prototype of the Cepheid class of short-period variables (see Variable Stars, pages 114-115). Its variability was discovered by the English astronomer John Goodricke in 1784. Delta's brightness fluctuates between magnitudes 3·6 and 4·3 over a well-determined period of 5·36634 days. Magnitude 3·35 Zeta and 4·19 Epsilon are useful comparison stars. Delta Cephei is also a double star; a magnitude 6·3 companion is located 41″ from Delta itself.

M52 OPEN CLUSTER IN CASSIOPEIA

Cassiopeia contains a number of open star-clusters within the reach of binoculars or a small telescope, one of the brightest of which is M52 (NGC 7654) seen in the accompanying photograph (*below*). This collection of around a hundred stars shines with an overall magnitude of 6·9 from a distance of over 5,000 light years. Located immediately to the south of the star 4 Cassiopeiae, M52 is easily found by following the train of stars away from Beta Cassiopeiae as shown on the finder chart (*below*).

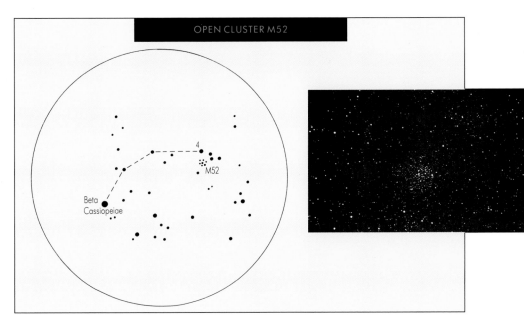

OPEN CLUSTER M52

73

DOUBLE AND MULTIPLE STARS

DOUBLE STARS

Double stars are of two types. Optical doubles consist of two stars which happen to lie more or less in the same line of sight as seen from Earth and which therefore only appear to lie together. In reality, one star may lie many times further away than its 'companion' as can be seen in the accompanying diagram (*right*).

Binary systems, on the other hand, are made up of two stars which are physically associated. The members of binary systems do not actually orbit each other. Rather, they orbit their common centre of gravity. To put it simply, the centre of gravity of a binary star, or indeed of any system (such as the Earth and Moon), is the point at which the two objects would balance if joined together by a bar. For example, if you attach unequal weights to either end of a bar,

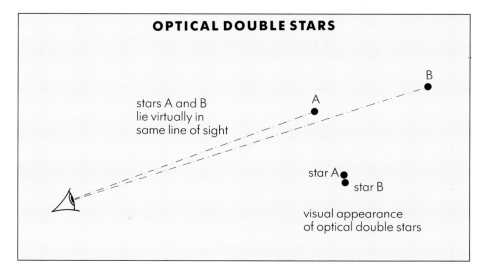

OPTICAL DOUBLE STARS

stars A and B lie virtually in same line of sight

star A
star B

visual appearance of optical double stars

their balance point would be located nearer the heavier weight. Or, with two people on a see-saw, the heavier of the two will have to sit closer to the fulcrum in order for both to balance properly. The greater the weight difference, the

closer to the fulcrum the heavier person needs to sit. In the case of the Earth-Moon system, the difference in mass between the two objects is so great that their centre of gravity actually lies within the sphere of the Earth.

MULTIPLE STARS

Some systems comprise three or more stars, a classic example being the sextuple star Castor in Gemini (see page 83). A multiple system can also be seen in the Plough. To anyone with keen eyesight Mizar, the central star in the handle of the Plough, appears to be unusual. A careful look will show that Mizar is accompanied by the 4th magnitude star Alcor. To complicate matters further, Mizar is itself a close double (resolvable only through telescopes), and a fourth star, Sidus Ludovicianum, makes up what is a stellar quartet. This photograph (*above right*) shows Alcor (the fainter of the bright pair at centre of picture) with Mizar (the brightest star). Although the close companion to Mizar isn't shown here, Sidus Ludovicianum can be seen below the line joining Alcor and Mizar and forming a triangle with this pair. Further details of this system are given on pages 70-71.

The orbital motions of such multiple stars can be quite complicated. In a triple star system, for example, we may have a

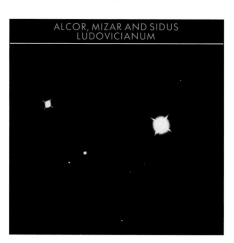

ALCOR, MIZAR AND SIDUS LUDOVICIANUM

binary star with components orbiting their common centre of gravity. A third component would then travel round the common centre of gravity between it and the orbiting pair as a whole! Sir William Herschel recognized the true nature of binary stars almost two centuries ago when he noticed that the positions of the stars in double systems changed relative to each other over periods of time. He deduced that in each case, the stars he had observed were actually orbiting each other.

DESCRIBING THE VISUAL APPEARANCE OF DOUBLE STARS

When describing the visual appearance of a double star, the magnitudes of the two components are given, followed by the angular distance between them. This in turn is often followed by the Position Angle (PA), this being the angle measured through east from an imaginary line extending north from the brighter component (see diagram A *opposite*). Expression of the PA can give the observer an idea of the visual appearance of a binary prior to observation.

The very nature of a binary system means that its separation and PA are subject to change. The apparent motion of the fainter star relative to the brighter star can be either direct or retrograde (see diagram B *opposite*), producing either an increase or reduction in the Position Angle. Changes in separation arise through a combination of their actual distance apart and the angle from which we view the binary.

ALBIREO

Many double stars offer beautiful colour contrasts between their components, including the marvellous Albireo (Beta Cygni) pictured here (*right*). Appearing as a 3rd magnitude star to the unaided eye, telescopes reveal that yellowish Albireo has a 5th magnitude sapphire component located 34·3″ away. The pair can be resolved in virtually any small telescope and even good binoculars will split the two components.

No orbital motion has been detected between the two stars in the Albireo system, although they are believed to be associated with each other and to be

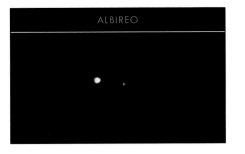
ALBIREO

travelling together through space. Their actual separation is in excess of 645,000 million km (400,000 million miles), well over 4,000 times the distance between the Earth and Sun.

DESCRIBING DOUBLE STARS

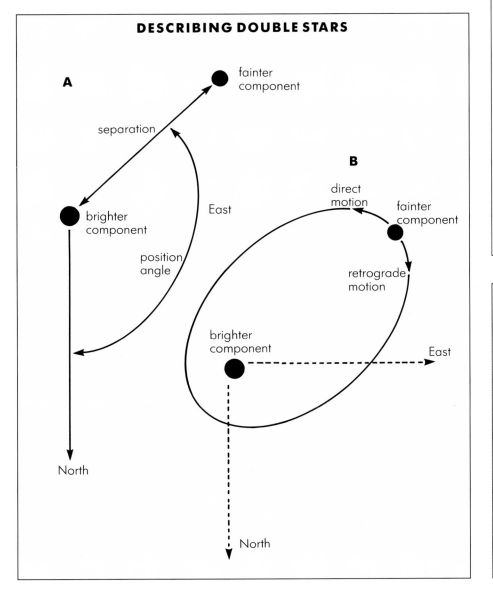

A
fainter component
separation
brighter component
East
position angle
North

B
direct motion
fainter component
retrograde motion
brighter component
East
North

ANGULAR MEASUREMENT

In order to express distances between points on the celestial sphere, astronomers use angular measurement. Basically, angular measurement expresses the angle that would be subtended at the eye by two lines, each of which is lined up with one of the two points in the sky. For example, the angle subtended by Merak and Dubhe, the two stars at the end of the Plough, is about 5°, as shown in this diagram. In other words, we say that Merak and Dubhe are 5° apart.

Expressing angular distances in terms of whole degrees is sufficient when we are discussing large areas of sky. However, in order to express much smaller angular separations, such as the apparent diameter of a planet or the distance between two stars in a double system, much smaller units must be used. In the same way as large units of distance, such as kilometres, are split into smaller metres and centimetres, whole degrees are subdivided into smaller units. A degree is divided into 60 minutes (often referred to as minutes of arc and distinguished by the symbol ′), each minute of arc being further divided into 60 seconds of arc (distinguished by the symbol ″). It can be seen that a degree is made up of 60 × 60 or 3,600″ and it is clear that angles expressed in seconds of arc must be very small indeed.

ANGULAR MEASUREMENT OF MERAK AND DUBHE

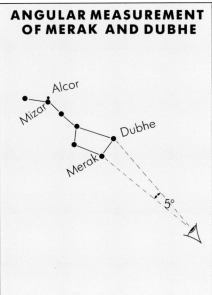

Mizar
Alcor
Dubhe
Merak
5°

MAIN CHART OF NORTHERN WINTER/ SOUTHERN SUMMER STARS

For observers across the globe, Orion is the centrepiece of the sky during northern winter and southern summer. The outline of Orion is formed from the four stars Betelgeuse, Bellatrix, Rigel and Saiph, while at their centre is the close trio of stars forming the Belt of Orion. Extending the line formed by these three stars to the north-west brings us to the Hyades and Pleiades open star clusters in Taurus (the Bull). Following the line in the opposite direction takes us to Sirius in Canis Major (the Great Dog), the brightest star in the sky. Canopus, the second brightest star, is also seen on this chart. Canopus is the leading member of Carina (the Keel) and, although lost to observers in the United Kingdom, can be glimpsed from the southern United States. Old star charts depict the large constellation Argo Navis (the Ship Argo). This group was split up into the four separate constellations Carina, Puppis (the Stern), Vela (the Sail) and Pyxis (the Mariner's Compass) which grace modern star charts.

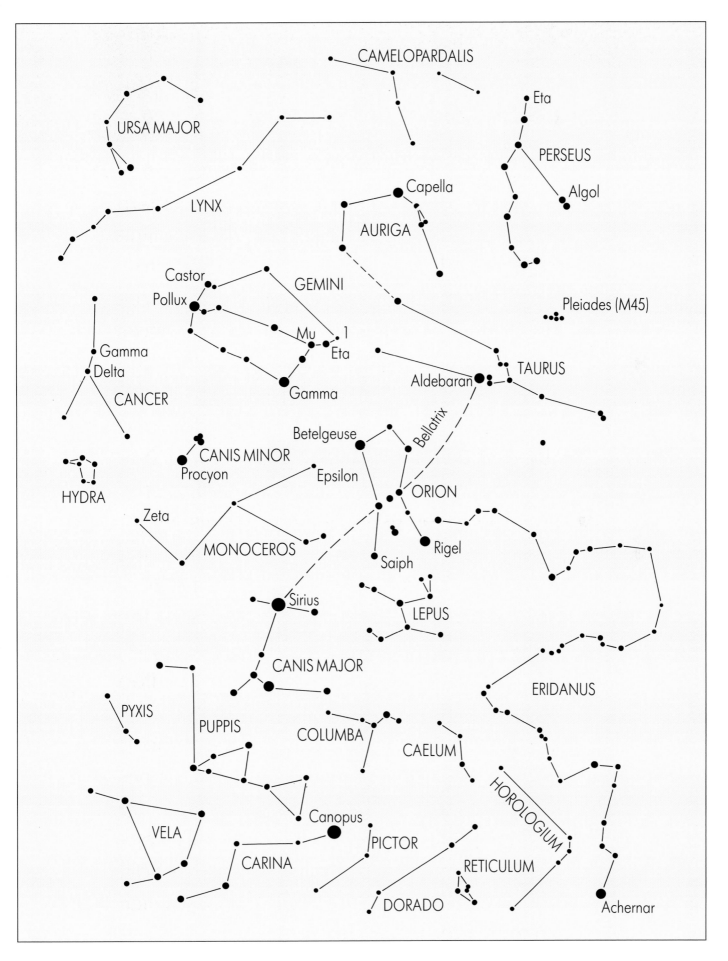

AURIGA

The distinctively shaped constellation of Auriga (see chart *below*) is easily located to the north of Orion, situated almost overhead for observers at mid-northern latitudes during winter evenings. Its brightest star, the golden-yellow Capella, is unmistakable. Shining with a magnitude of 0·06, Capella is the sixth brightest star in the sky. It lies at a distance of 45 light years and has a true luminosity equal to around 160 times that of the Sun. The name Capella is derived from the Latin word for 'she-goat'. One legend that has come down to us refers to a Cretan goat named Amaltheia who suckled Jupiter and who was repaid for her efforts by being placed among the stars. Alongside Capella are her two kids, known collectively by their Latin name Haedi and represented by the two stars Eta and Zeta Aurigae.

At first sight the main stars of Auriga seem to form a neat pentagon, yet Beta Tauri – the southernmost star in the pentagon – is actually a member of the neighbouring constellation Taurus. Lying on the boundary between Auriga and Taurus, this star was originally known as Gamma Aurigae, but since its transfer has been designated Beta Tauri.

Auriga lies across the Milky Way and contains many rich star-fields. Binoculars will reveal the pearly glow of the open star clusters M37 (NGC 2099), M36 (NGC 1960) and M38 (NGC 1912) which form a neat trio crossing the line between Theta Aurigae and Beta Tauri. Carefully sweeping the area with binoculars will enable you to pick out all three clusters, and 7 × 50 binoculars will bring them all into the same field of view. However, resolution of individual member stars is only possible through telescopes of at least 60-mm (2·5-in) aperture.

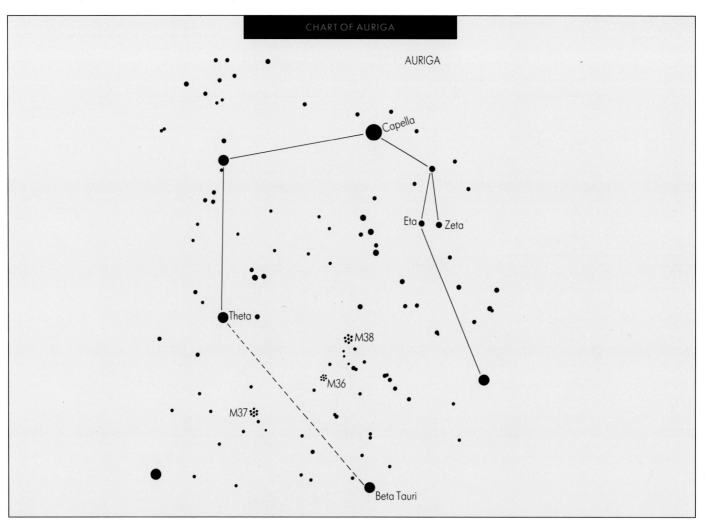

CHART OF AURIGA

AURIGA

Capella

Eta Zeta

Theta

M38

M36

M37

Beta Tauri

OPEN CLUSTER M37

At magnitude 5·6, M37 (see *right*) is the brightest of the trio and is dimly visible to the naked eye under really dark, clear skies. This superb object lies at a distance of 4,600 light years and is thought to contain up to 500 stars within a volume of space some 25 light years in diameter. Large telescopes will bring out between 150 and 200 individual starlike points within the cluster. Discovered by Charles Messier in 1764, M37 is considered to be the finest of Auriga's clusters. The English observer Rev Thomas William Webb drew attention to this object, referring to the whole field '. . . being strewed . . . with sparkling gold-dust; extremely beautiful, one of the finest of its class'.

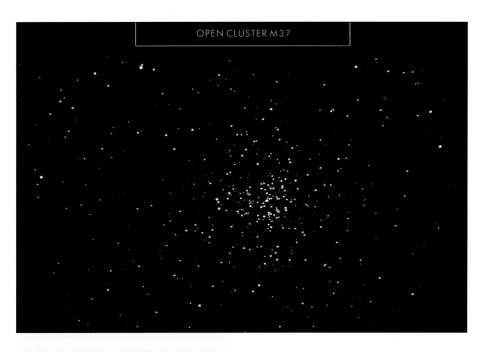

OPEN CLUSTER M37

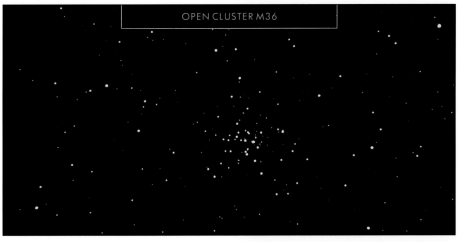

OPEN CLUSTER M36

OPEN CLUSTER M36

The nearby open cluster M36 (see *left*) lies around 5° to the south-west of Theta Aurigae and contains around 60 stars shining with a combined magnitude of 6·0. This stellar gathering lies at a distance of some 4,000 light years and has an overall diameter in the order of 14 light years. Described by Webb as a 'Beautiful assemblage of stars . . . very regularly arranged', M36 displays a bright condensation of stars when viewed through small telescopes, larger instruments resolving outlying members.

OPEN CLUSTER M38

At magnitude 6·4, M38 (see *right*) is the faintest of the three clusters described here. Shining from a distance of 4,200 light years, the 100 or so stars that form M38 are spread out over a volume of space some 25 light years across, making it comparable in size to M37. Best views are obtained with wide-field telescopes which will show M38 against the backdrop of Milky Way star fields.

OPEN CLUSTER M38

ORION

A part from the Plough, Orion is undoubtedly the best-known and most easily recognized group of stars in the entire sky. Orion is visible from every populated region of the Earth, and its conspicuous rectangular shape, containing the close trio of stars marking Orion's Belt, is unmistakable.

BETELGEUSE AND RIGEL

Orion's two main stars are the red giant Betelgeuse (Alpha Orionis) and the blue-white supergiant Rigel (Beta Orionis) which, shining at magnitudes 0·7 and 0·14, are the 11th and 7th brightest stars in the sky. Betelgeuse, whose name is derived from the Arabic for 'Arm of the Central One', is in fact variable, its brightness fluctuating by up to half-a-magnitude either way over a period of around 5·7 years. It is one of the largest known stars.

Rigel is a brilliant star — one of the most luminous objects in the Galaxy. Its name is derived from the Arabic for 'Left Leg of the Giant' and its true luminosity is thought to be approaching 60,000 times that of our Sun. Rigel's diameter is estimated to be some 50 times that of our star. Rigel has a magnitude 6·7 companion located 9" away; easily swamped by the glare from Rigel, telescopes of at least 100-mm (4-in) aperture are required to glimpse it.

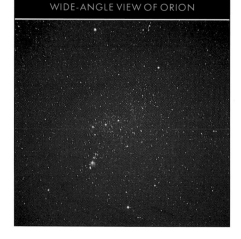
WIDE-ANGLE VIEW OF ORION

ORION NEBULA

The magnificent Orion Nebula (M42) is located a little way to the south of the three stars marking the Belt of Orion, where it can be seen as a shimmering nebulous patch of light with the naked eye. Located at a distance of around 1,600 light years, its light is produced through the effects of ultra-violet radiation from the multiple star Theta Orionis, located at the heart of M42 (*left*).

ORION NEBULA

ORION AND THE HYADES

Orion is a marvellous pointer to other stars and constellations. This photograph of Orion and part of neighbouring Taurus (*right*) shows how a line extended through Orion's Belt towards the upper right (north-west) leads to the bright star Aldebaran and the adjoining Hyades open cluster. (Also see the main chart on pages 76-77.) Aldebaran and the Hyades, together with other objects of interest in Taurus, are covered in more detail on page 82.

ORION AND THE HYADES

ERIDANUS

The long, winding shape of Eridanus (see chart *below*) stretches from near a point just to the west of Rigel (Beta Orionis) to the west and then down in a meandering path, eventually reaching Achernar, the brightest star in the constellation. Observers at mid-northern latitudes will get their best views of the group during mid- to late-evenings in November, at which time you will be able to make out the member stars as far south as Upsilon$_1$, provided the sky immediately above the southern horizon is really clear. Stars below this will only come into view from more southerly latitudes; Achernar itself is only observable from locations below around latitude 30°N.

Achernar is a hot blue giant-star shining from a distance of around 120 light years. It is the 9th brightest star in the sky, its true luminosity being well over 600 times that of the Sun. Its diameter is around 9·5 million km (6 million miles) and the surface temperature in the region of 14,000°C (25,000°F).

Unlike many other constellations, Eridanus resembles the object that it is supposed to represent, although it has to be said that you cannot really go wrong when trying to form a pattern of a river out of stars! There are several double stars accessible to small telescopes within the constellation, notably the pretty object 32 Eridani, whose yellow and blue magnitude 5·0 and 6·3 components are separated by 6·9". A 60-mm (2½-in) telescope should resolve this pair. The same instrument will split the magnitude 4·9 and 5·4 members of f Eridani. These two stars lie 7·9" apart. Theta Eridani has white components of magnitude 3·4 and magnitude 4·4 separated by 8·2".

Of particular interest is the triple star Omicron$_2$ Eridani which lies at a distance of only 16 light years. The main component has a magnitude of 4·5 star, 82·8" away from which is a magnitude 9·5 white dwarf. This is the most easily observed white dwarf star in the sky and is a must for owners of small telescopes. Instruments of 100 mm (4 in) or more should bring out this pair, which were first resolved by William Herschel in 1783. However, the white dwarf companion is itself a binary star with a magnitude 10·8 red dwarf star in a 248-year orbit around it. The current separation of the pair is currently near its maximum of 9".

THE 20 BRIGHTEST STARS				
NAME	CONSTELLATION	APPARENT MAGNITUDE	ABSOLUTE MAGNITUDE	DISTANCE (L.Y.)
SIRIUS	CANIS MAJOR	−1·42	+1·4	8·8
CANOPUS	CARINA	−0·72	−8·5	1,170
ALPHA CENTAURI	CENTAURUS	−0·27	+4·4	4·24
ARCTURUS	BOOTES	−0·06	−0·2	37
VEGA	LYRA	+0·04	+0·5	27
CAPELLA	AURIGA	0·06	−0·3	45
RIGEL	ORION	0·14	−7·1	900
PROCYON	CANIS MINOR	0·35	+2·6	11·3
ACHERNAR	ERIDANUS	0·53	−1·6	120
HADAR	CENTAURUS	0·66	−5·1	490
BETELGEUSE	ORION	0·70 (V)	−5·6 (V)	650
ALTAIR	AQUILA	0·77	+2·2	16
ALDEBARAN	TAURUS	0·86	−0·3	68
ACRUX	CRUX	0·87	−3·9	370
ANTARES	SCORPIUS	0·92 (V)	−4·7 (V)	520
SPICA	VIRGO	1·00 (V)	−3·5 (V)	275
POLLUX	GEMINI	1·16	+0·2	35
FOMALHAUT	PISCES AUSTRINUS	1·17	+2·0	23
DENEB	CYGNUS	1·26	−7·5	1,600
BETA CRUCIS	CRUX	1·28	−4·6	490

NOTE: (V) DENOTES STAR IS VARIABLE

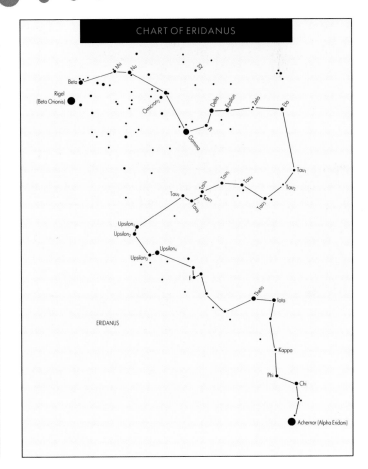

CHART OF ERIDANUS

HYADES & PLEIADES OPEN CLUSTERS

Taurus (the Bull) plays host to two of the best-known open star clusters in the entire sky, both of which are seen in this photograph (*below*) and accompanying chart (*below right*). The prominent V-shape of the widely-scattered Hyades cluster stands out to the lower left of centre while the much more compact Pleiades cluster (M45) is seen to its northwest (upper right). Aldebaran, the brightest star in Taurus, can be seen here as a conspicuous reddish object at the end of the lower arm of the 'V'.

Aldebaran is not actually a member of the cluster, its distance of 68 light years putting it only half-way between ourselves and the Hyades.

The Hyades covers an area of sky over 5° across, slightly larger than the distance between Merak and Dubhe in the Plough (see page 70). The group is well worth sweeping with binoculars or a rich-field telescope; several double stars are within the grasp of small instruments. The 100 or so cluster members visible with small telescopes display

numerous colours and amply reward time spent observing the area.

Located around 12° to the northwest of the Hyades is the Pleiades cluster, a collection of young, hot stars located at a distance of just over 400 light years. The Pleiades covers an area of sky some 2° across and is best observed through either binoculars or a rich-field telescope, the wide fields of view enabling you to see the entire group. Further details of both the Hyades and Pleiades can be found on pages 124-125.

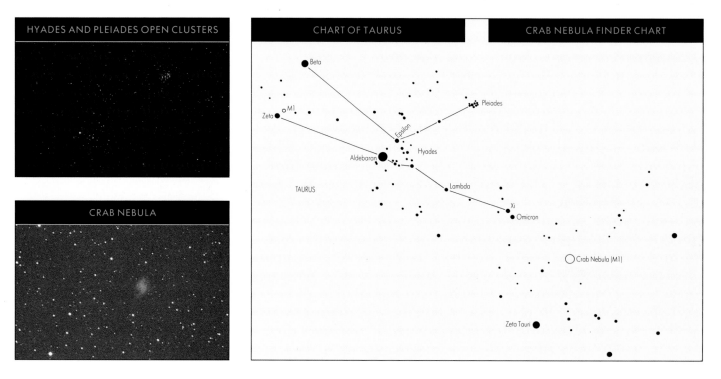

HYADES AND PLEIADES OPEN CLUSTERS

CRAB NEBULA

CHART OF TAURUS

CRAB NEBULA FINDER CHART

THE CRAB NEBULA

The chart of Taurus (see *above right*) shows the location of the star Zeta Tauri, from which the Crab Nebula (M1) can be located using the accompanying finder chart (*above right*) using Zeta Tauri as your guide star.

The Crab Nebula is the best-known

example of a supernova remnant (see Stellar Evolution, pages 60-63) and represents the scattered remains of a supernova seen to explode in July 1054. The actual distance of M1 is around 6,000 light years, which puts the actual date of the explosion at around 5,000 BC – the

light from the supernova took 60 centuries to reach us! The overall magnitude of the Crab Nebula is 8·5 making it dimly visible as a greyish, oval patch of light through binoculars or a small telescope (see *above*). Dark, clear, moonless skies are essential to see this object.

CANIS MINOR AND GEMINI

This wide-field view (see *below*) shows Canis Minor (the Little Dog), and Gemini (the Twins). These constellations can be identified from the main whole-sky chart on pages 76-77.

Canis Minor is one of the smallest constellations, its leading star Procyon, at magnitude 0·35 the eighth brightest star in the sky, being located almost half-way up the left margin of picture. Procyon is also the 14th nearest star to us, shining from a distance of just 11·3 light years. (A list of the 20 nearest stars is included on these pages.) Procyon is actually a binary star; its two components orbiting each other every 40·65 years. Their angular separation varies between 2·2" and 5·0", although the 11th magnitude companion is extremely difficult to see in the glare of the brilliant primary. This companion, known as Procyon B, is a white dwarf star (see Stellar Evolution, pages 60-63) with an estimated diameter of only around 27,000 km (17,000 miles) but a mass equal to around 65 per cent that of the Sun. Given these figures, the density of Procyon B works out to be around 125 kg/cm^3 (2 tonnes/in^3)!

Just above and to the right of centre is the prominent pair of stars Castor (right) and Pollux, the leading stars of Gemini and the 23rd and 17th brightest stars in the sky respectively. From this pair, the rest of Gemini can be seen extending towards the west (lower margin) of picture. Castor and Pollux form a conspicuous pair, lying just 4·5° apart. Telescopes of 100-mm (4-in) aperture will show that Castor is a binary star with components of magnitudes 2·0 and 2·85, currently separated by 2·5". The angular distance between these two stars, known as Castor A and Castor B, varies between 1·8" and 5·0" over an orbital period of around 400 years. A third star, magnitude 9·1 Castor C, has been found 72·5" away and to orbit the main pair every 100 centuries or so. To complicate matters further, each of these stars has been found to be a spectroscopic binary (see Double and Multiple Stars, page 74–75), making Castor a sextuple star system!

Pollux is somewhat uninteresting in comparison! Shining with a magnitude of 1·16, this yellowish star has a true luminosity equal to around 35 times that of the Sun. Its surface temperature is 4,500°C (8,100°F) and its diameter over ten times that of our star.

THE 20 NEAREST STARS

NAME	CONSTELLATION	APPARENT MAGNITUDE	DISTANCE (L.Y.)
PROXIMA CENTAURI	CENTAURUS	+10·7	4·3
ALPHA CENTAURI*	CENTAURUS	− 0·27	4·34
BARNARD'S STAR	OPHIUCHUS	+ 9·5	6·0
WOLF 359	LEO	+13·6	7·7
LALANDE 21185	URSA MAJOR	+ 7·6	8·3
SIRIUS*	CANIS MAJOR	− 1·42	8·7
LUYTEN 726-8*	CETUS	+12·3	9·0
ROSS 154	SAGITTARIUS	+10·6	9·5
ROSS 248	ANDROMEDA	+12·2	10·3
EPSILON ERIDANI	ERIDANUS	+ 3·7	10·8
LUYTEN 789-6	AQUARIUS	+12·2	10·8
ROSS 128	VIRGO	+11·1	10·9
61 CYGNI*	CYGNUS	+ 5·2	11·1
PROCYON*	CANIS MINOR	+ 0·35	11·3
EPSILON INDI	INDUS	+ 4·7	11·4
SIGMA 2398*	DRACO	+ 8·8	11·4
GROOMBRIDGE 34*	ANDROMEDA	+ 8·0	11·7
TAU CETI	CETUS	+ 3·5	11·8
LACAILLE 9352	PISCES AUSTRINUS	+ 7·4	11·9
BD +5°1668	CANIS MINOR	+ 9·8	12·3

*DENOTES STAR HAS A COMPANION

CANIS MINOR AND GEMINI

NEBULAE

Nebulae are concentrated patches of rarified dust and gas, spread throughout interstellar space. Observed in other galaxies as well as our own, many examples have been catalogued. Their collective name is derived from the Latin word *nebula* meaning 'mist' or 'cloud', and there are three basic types. Emission nebulae emit their own light glowing through the effects of young, hot stars embedded within them, the energy from these stars causing the gas in the nebula to fluoresce. Reflection nebulae, on the other hand, contain no stars hot enough to cause the gas within them to shine, the dust particles within them merely reflecting the light from nearby stars. Dark nebulae, as their name suggests, contain no illuminating stars and appear as black patches against brighter backgrounds.

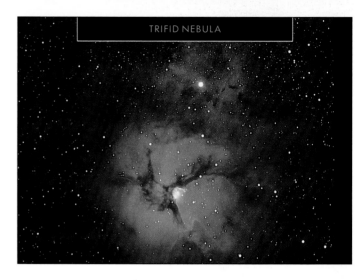
TRIFID NEBULA

EMISSION NEBULAE

The characteristic red colour of emission nebulae, such as the Trifid Nebula, seen here (*above right*), comes about through the absorption of strong ultraviolet energy from very hot young stars. These stars, embedded deep within the nebula, cause the hydrogen atoms to become ionized, during which process they lose their electrons, causing the hydrogen to glow at visible wavelengths.

Nebulae of this type are also known as HII regions, the Trifid Nebula being one of the most famous examples. Its name was probably first coined by the English astronomer John Herschel who described it as '. . . consisting of 3 bright and irregularly formed nebulous masses . . . surround[ing] a sort of 3-forked rift or vacant area . . . quite void of nebulous light . . .'. The dark rifts are contained within the reddish southern portion of the nebula, and are seen to radiate outwards from a central triple star. This star is clearly visible at the inner corner of one of the nebula's three bright regions. The overall visual effect of the Trifid Nebula is quite stunning, the dark lanes beautifully trisecting the bright HII region beyond.

Just to the north of the main HII region is a patch of blue nebulosity. This is a cloud of dust and gas which reveals itself through the reflection of starlight from the dust grains embedded within it. The illuminating stars of reflection nebulae such as the one seen here are not hot enough to cause the ionization processes characteristic of emission nebulae.

REFLECTION NEBULAE

The main photograph (*below*) shows regions of reflection nebulosity swirling around the central stars of the Pleiades open cluster in Taurus. The Pleiades is a young cluster, its stars probably forming only around 20 million years or so ago. The associated nebulosity envelopes the entire cluster and is thought to be the remnant of the original nebula from which the stars of the Pleiades were formed. It is difficult to see, but a significant amount of detail is revealed on long-exposure photographs. The Pleiades nebulosity is probably the best-known example of reflection nebulosity in the heavens.

The brightest portion of the Pleiades nebulosity is that which surrounds Merope. Its visual appearance, seen in this close-up photograph (see *opposite page, top*), consists of wisps and streamers, and has been likened to cirrus clouds here on Earth. Similar structuring is observed elsewhere in the Pleiades nebulosity.

PART OF PLEIADES OPEN CLUSTER

MEROPE IN PLEIADES

DARK NEBULAE

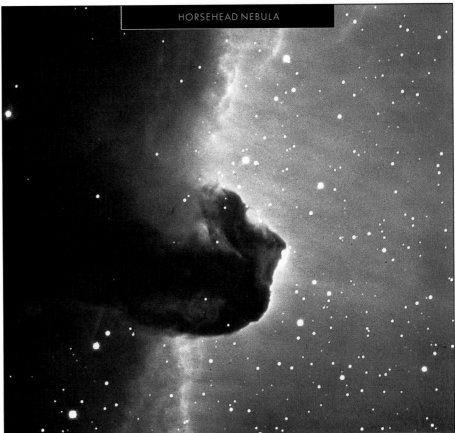

HORSEHEAD NEBULA

Located half a degree to the south of the bright star Zeta Orionis, the Horsehead Nebula is typical of dark nebulae, appearing as a dark patch seen against a brighter background as seen in this photograph (*right*). In the case of the Horsehead Nebula, the backdrop is provided by the emission nebula IC434. Although many dark nebulae have been catalogued, the Horsehead is certainly the best known and is widely regarded as being the classic nebula of its type. Dark nebulae are essentially like reflection nebulae, although they contain no illuminating stars. The Horsehead is itself simply an extension of a much larger dust cloud situated to the east. It lies at a distance of 1,200 light years and is thought to be one light year in diameter.

The Horsehead Nebula was discovered photographically on a plate taken in 1889 by the American astronomer Edward Charles Pickering. It is extremely difficult to see, visually, and long-exposure photographs are required to bring out its intricate details. At first it was thought that the object was nothing more than a gap roughly halfway along the eastern edge of the brighter emission nebula. It was the American astronomer Edward Emerson Barnard who realized that the Horsehead and other similar objects, which had been thought of as dark voids in the Milky Way, were in fact dark patches of interstellar dust and gas.

Barnard carried out a survey of dark nebulae, and in 1927 published his *Photographic Atlas of Selected Regions of the Milky Way*. This contained many examples of dark nebulae, to each of which he assigned numbers. Many dark nebulae are now also referred to by their Barnard numbers, including the Horsehead Nebula, or B33.

The seven stars forming the Plough are located well up in the sky for northern hemisphere observers. These stars are merely the brightest in the large constellation Ursa Major (the Great Bear). Following the curve of the three stars in the Plough handle, as shown, brings us to Arcturus, the leading star of Bootes (the Herdsman) and Spica, brightest member of Virgo (the Virgin). Crossing this chart can be seen the winding form of Hydra (the Water Snake), the largest constellation in the sky. The Head of Hydra is shown on the chart of Northern Winter/Southern Summer stars (see pages 76/77). To the south of Hydra are the conspicuous constellations Centaurus (the Centaur) and Crux (the Cross – often referred to as the Southern Cross). Deceiving the first-time observer are the four stars Kappa and Delta Velorum and Iota and Epsilon Carinae which together form a pattern similar to that of Crux. However, the stars in the 'False Cross' are somewhat fainter than those in Crux.

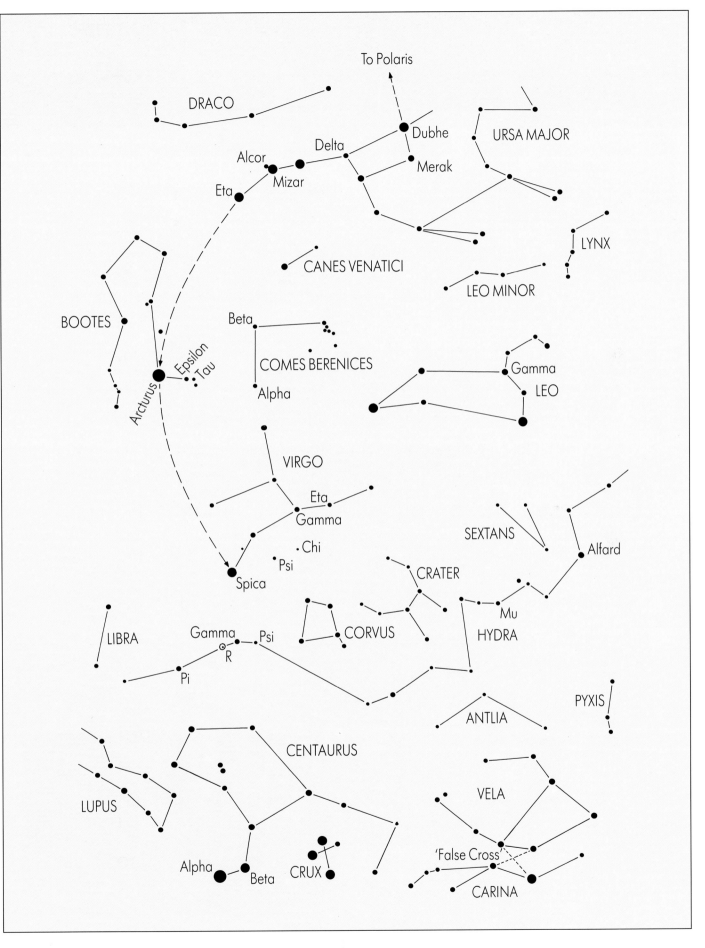

VIRGO

The constellation Virgo contains a large number of galaxies, a number of which are within the light grasp of moderate telescopes. This area of sky contains the Virgo Cluster of Galaxies (see pages 116-117), the huge swarm of island universes extending upwards into the constellation of Coma Berenices.

GALAXY M49

One of the brightest of the Virgo galaxies is the nearly-spherical elliptical galaxy M49 (NGC 4472) (see *below inset*). This can be picked up by first locating the star Delta Virginis (details of how to locate Virgo and its leading stars can be found in Seek Out A Quasar!, pages 94-95). Once you have spotted Delta, look for the circlet of fainter stars to its north-west, shown on the accompany-ing finder chart (*below*). M49 will be seen nestled between two 6th magnitude stars in the circlet.

Shining from a distance of just over 40 million light years, M49 is one of the brightest galaxies in the Virgo Cluster. With a diameter of around 50,000 light years and a mass estimated to be around five times that of the entire Milky Way Galaxy, it is also one of the largest and most massive elliptical galaxies known. M49 was discovered by the French astronomer Charles Messier in 1771 and is dimly visible in large binoculars as an almost-circular nebulous patch of light, providing the sky is dark and clear. Telescopes of 100-mm (4-in) aperture will show its bright centre, the surrounding parts of the galaxy being seen to fade towards the edges.

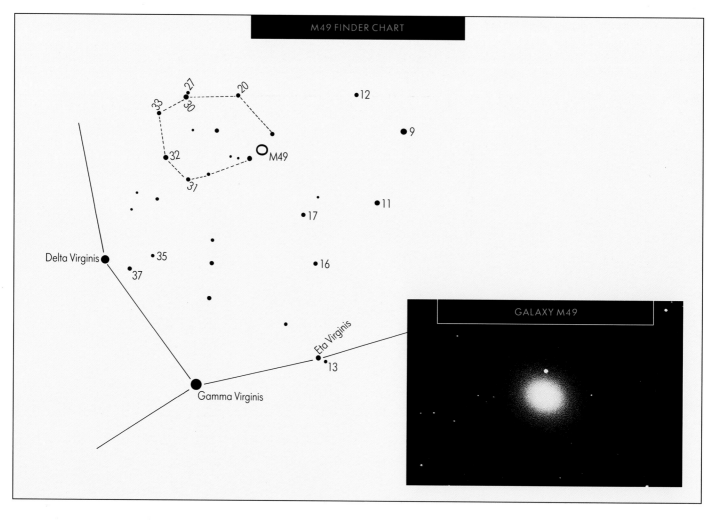

M49 FINDER CHART

GALAXY M49

HYDRA

ydra (the Water Snake) is the largest constellation in the sky and its long and winding form, running beneath the constellations Corvus, Crater and Sextans, can be seen from most areas of the world. The brightest star in Hydra is Alfard (Alpha Hydrae) with a magnitude of 1·98.

GALAXY M83

The 'Southern Pinwheel' galaxy M83 (NGC 5236) can be found in the starfield to the south-west of Pi Hydrae (see main chart of Hydra *below*) and can be located using the guide stars R, Gamma and Psi Hydrae (see accompanying finder chart *below*). Located at a distance of around 10 million light years and lying face-on to us, 8th magnitude M83 has a diameter of 30,000 light years. It is one of the finest objects of its type in the sky and can be spotted as a fairly conspicuous nebulous disc in binoculars. Small telescopes will show that it is brighter at its centre, but larger instruments of 200 mm (8 in) or more bring out some detail of the structure within the spiral arms (see *right*). Its angular diameter is around 10', roughly a third that of the Moon.

GALAXY M83

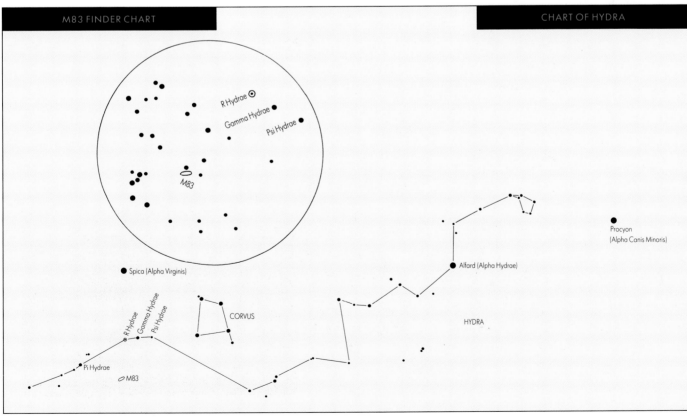

M83 FINDER CHART

CHART OF HYDRA

R Hydrae

Gamma Hydrae

Psi Hydrae

M83

Spica (Alpha Virginis)

R Hydrae

Gamma Hydrae

Psi Hydrae

CORVUS

Pi Hydrae

M83

Alfard (Alpha Hydrae)

HYDRA

Procyon
(Alpha Canis Minoris)

GEMINI AND M35 OPEN CLUSTER

The constellation Gemini contains one of the most splendid examples of an open cluster in the sky. M35 (NGC 2168) lies just to the north of the line of stars comprising Mu, Eta, 6, 4, 3 and 1 Geminorum, as shown on the accompanying finder chart (*right*). Shining at magnitude 5·5 it can be easily detected using binoculars, and can even be glimpsed with the naked eye under good seeing conditions. Its 120 member stars span an area of sky roughly half-a-degree across, making it comparable in size to the full lunar disc.

The photograph (*below*) shows M35, visible at centre, with the stars shown in the finder chart. Particularly prominent here is the line of stars from Mu through to 1 Geminorum, detailed above. M35 lies at a distance of around 2,200 light years and has a true diameter of 30 light years. When viewed through binoculars, M35 appears as a nebulous patch of light with some individual stars resolved. Telescopes of 200-mm (8-in) aperture or more reveals a field full of stars, arranged in lines and curves, making M35 a splendid telescopic object.

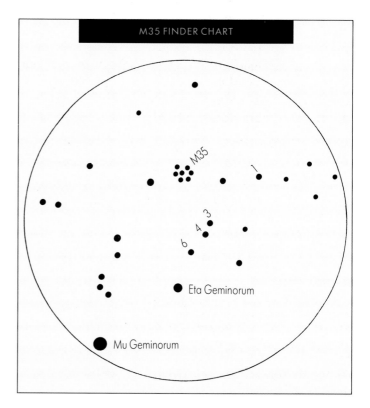

M35 FINDER CHART

M35 OPEN CLUSTER

CANCER

The right-hand (western) half of the wide-field photograph (*right*) includes the approximate area covered on the accompanying chart of Cancer (*below right*). At upper right (north west) of both is the prominent pair of stars Castor and Pollux (Alpha and Beta Geminorum) while the quadrilateral of stars formed from Gamma, Delta, Theta and Eta Cancri is visible near the bottom margin of the photograph, just to the right of centre. By cross-referencing between the photo and chart, several more stars can be identified.

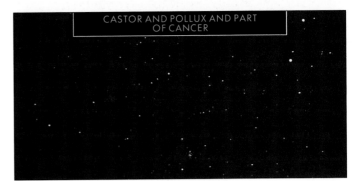

CASTOR AND POLLUX AND PART OF CANCER

M44 AND M67 OPEN CLUSTERS

Cancer (the Crab) hosts two famous open clusters. The brightest and best-known of these is M44 (NGC 2632), also known as 'Praesepe' or the 'Beehive'. This cluster is located within the quadrilateral of stars described above and shown on the accompanying photograph (*bottom right*). It shines from a distance of around 500 light years. The overall magnitude of M44 is 3·1, making it visible to the naked eye. It was first recorded by the Greek astronomer Aratus in around 260 BC. This sprawling group of around 350 or so stars has a diameter of just over three times that of a full lunar disc although it contains only a few relatively bright stars. The best views of M44 are obtained using wide-field telescopes or binoculars; 7 × 50 binoculars take in the entire cluster and reveal many individual members.

To the south of M44 is the often-overlooked M67 (NGC 2682). This object lies within the line of stars formed from Acubens (Alpha Cancri), 60, 50 and 45 Cancri, shown on the chart of Cancer (*above right*). The overall magnitude of M67 is 6·9, which is well within the light grasp of binoculars, but below the limit of naked-eye visibility. Binoculars will show M67 as a small misty spot; resolution of individual member stars requires a telescope. Even small telescopes will partially resolve the cluster; 60-mm (2·5-in) instruments show a glowing patch of light surrounded by starlike points. Further information and a photograph of M67 will be found under Star Clusters on pages 124-125.

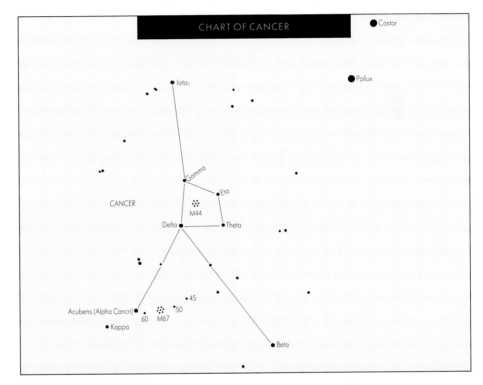

CHART OF CANCER

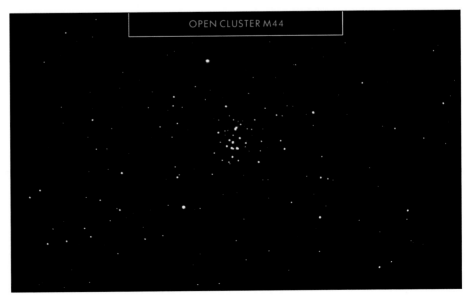

OPEN CLUSTER M44

CANES VENATICI

Although Canes Venatici is a small, somewhat obscure constellation, it does contain a wealth of deep-sky objects of interest to the backyard astronomer. Notable among these is the globular cluster M3. However, by far the most common type of object to grace the constellation are galaxies. Most of those visible within the boundaries of Canes Venatici are members of the Virgo Supercluster, a colossal gathering of island universes of which our Galaxy and the Local Group of galaxies are members.

GLOBULAR CLUSTER M3

Binoculars will enable you to locate the globular cluster M3 (NGC 5272) in Canes Venatici which lies practically on the border between Canes Venatici and neighbouring Bootes (see accompanying chart *opposite top*). Find the bright star Arcturus by following the curve of the Plough handle down towards the south. Arcturus is the first really bright star you come across. From here, use your binoculars to locate the pair of stars 9 and 11 Bootis which lie to the north of Arcturus. Once you have found this pair, follow the accompanying finder chart (*below right*) to M3 which lies roughly on a line between 9 and 11 Bootis and Beta Comae Berenicis. All three constellations mentioned here — Canes Venatici, Bootes and Coma Berenices — can be identified from the main chart on pages 86-87.

M3 is a splendid object (see *right*) and is regarded as one of the best of its class. Discovered by Charles Messier in 1764, it lies at a distance of between 35,000 and 40,000 light years. Several tens of thousands of stars have been counted in M3, all of which are contained within a volume of space some 220 light years across. When viewed through small binoculars, M3 appears as a spherical blur; resolution of individual stars is possible using telescopes. Instruments of 100-mm (4-in) aperture will give partial resolution, while 152-mm (6-in) telescopes will bring out many cluster members. Apertures of 300 mm (12 in) or more will, under dark, clear skies, resolve stars in M3 down to the core.

GLOBULAR CLUSTER M3

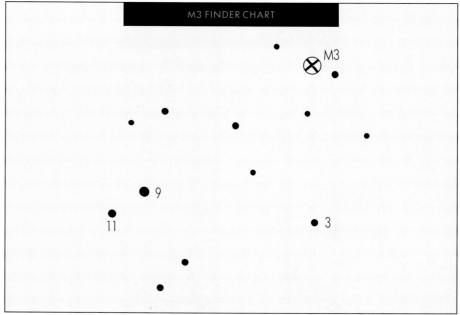

M3 FINDER CHART

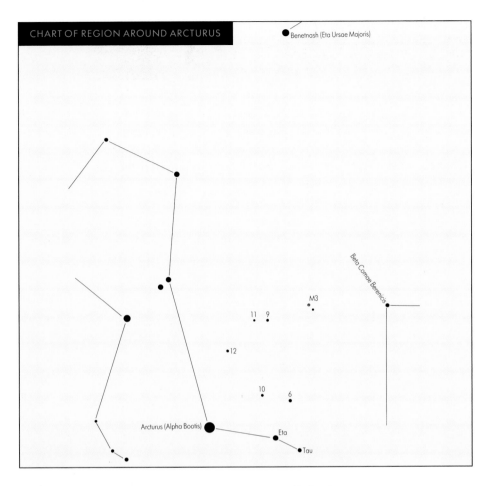

Benetnash (Eta Ursae Majoris)

Beta Comae Berenices

M3

11 9

12

10   6

Arcturus (Alpha Bootis)

Eta

Tau

GALAXY M51

One of the finest galaxies in this area, and indeed in the entire sky, is 8th magnitude M51 (NGC 5194), a face-on spiral system situated at a distance of around 35 million light years. See the accompanying finder chart for its location (*below*). Also known as the Whirlpool Galaxy, M51 was discovered by Messier in 1773 and has a diameter of around 100,000 light years. As the photograph shows (*below left*), M51 is accompanied by a satellite galaxy, NGC 5195, which is currently interacting with its larger companion. M51 is dimly visible using large binoculars, although dark, clear and moonless skies are essential. Telescopes of 60-mm (2·5-in) aperture will show the brighter central region of the galaxy, standing out against the surrounding, fainter glow of the spiral arms. Both M51 and NGC 5195 can be glimpsed through 100-mm (4-in) instruments, although the spiral nature of M51 requires telescopes of at least 254-mm (10-in) aperture to be resolvable, but only under first-rate observing conditions.

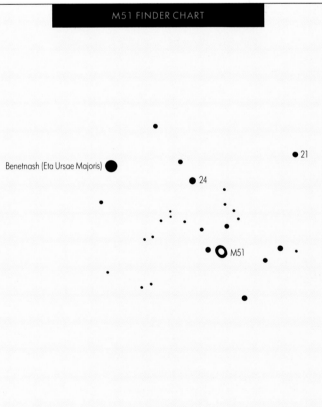

21

Benetnash (Eta Ursae Majoris)

24

M51

SEEK OUT A QUASAR!

Starlike in appearance, quasars (short for 'quasi-stellar radio sources') emit colossal amounts of energy, in some cases equivalent to tens or even hundreds of entire galaxies. However, the energy emitted by a typical quasar emerges from only a tiny volume of space comparable in size to our Solar System. Yet quasars lie at immense distances from us, the remotest examples being located well beyond the distance of the furthest galaxies. It is now thought that these enigmatic objects are extremely active galactic nuclei. Some astronomers believe that quasars may mark locations where material is falling into black holes.

Quasars were initially identified as strong radio-sources (although not all quasars are radio emitters) and were at first thought to lie within the Milky Way Galaxy. However, their extra-galactic status was revealed when the spectrum of quasar 3c273 was examined. This object was identified visually in 1963 when the optical component of a strong radio-source was identified by comparing its position with photographic plates taken at Palomar Observatory.

3c273 appeared as a faint bluish star although examination of its spectrum showed that it was anything but starlike. The spectral lines were seen to be greatly red-shifted which showed that 3c273 was moving away from us at very high speed. It followed that this object must lie well beyond the confines of our galaxy. Yet, since 3c273 can be seen in amateur telescopes means that it must have a very high luminosity.

Thousands of quasars have now been found. The most distant quasar identified to date is Q0000-26, which lies at a distance of around 13,000 million light years and which is receding at over 90 per cent of the speed of light! This immense distance means that the light being captured by astronomers from Q0000-26 set off towards us shortly after the formation of the Universe.

LOCATING QUASAR 3c273

Lying a few degrees north of the celestial equator, magnitude 12·8 3c273 can be seen in telescopes of 250-mm (10-in) aperture. The position of 3c273 in the sky, just over 4.5° to the north-west of Gamma Virginis, means that it is observable from all inhabited regions of the Earth. The best time for observation is during mid- to late-evening in April and May, at which time Virgo rides high in the southern/northern sky for northern/southern-hemisphere observers.

To track down 3c273, start by locating the constellation Virgo. For northern-hemisphere observers, its leading star Spica is found by following the curve of the Plough 'handle' down through Arcturus in Bootes and on as far again until Spica is reached. Southern-hemisphere observers can locate Virgo by using the prominent constellation Centaurus as a pointer; a line from Alpha through Iota Centauri, extended roughly 1½ times as far again, leads to Virgo. Finder Charts 1a and 1b, coupled with the main chart on pages 86-87, will enable you to find the constellation. Charts 1a and 1b, together with Chart 2

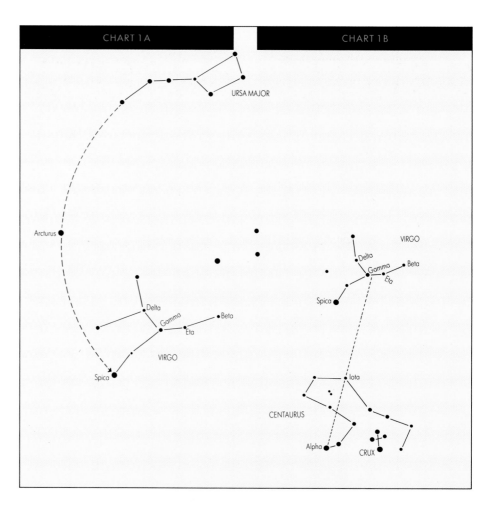

CHART 1A   CHART 1B

(described below) are drawn with north uppermost – showing the stars as seen in the northern hemisphere using either the naked eye or binoculars. Observers south of the Equator will need to make allowances for the fact that they will see these stars 'upside down' and not as shown here.

Once you have found Virgo, and its leading star Spica, pick out the triangle formed by the stars Gamma, Eta and 16 Virginis (see Chart 2). To the east of 16 Virginis extends a smaller triangle formed from 16 Virginis and the two 6th magnitude stars HD 109860 and HD 109896. Just to the south of *this* triangle is a smaller triangle formed by 7th magnitude ADS 8582 and 8th magnitude HD 108026 and HD 108228. This triangle is also shown in Chart 3, which has north to the bottom, so presenting the field of view as it would be seen through an astronomical telescope by northern-hemisphere observers.

Located immediately to the north of this is a fourth, even smaller triangle of 12th magnitude stars, found by extending the two curved lines of stars straddling the northern edge of the main triangle in Chart 3. Quasar 3c273 lies on the longest, northern edge of this tiny triangle, as shown in Chart 4, and is the brightest of the close pair of starlike points.

Although 250-mm (10in) or larger telescopes are needed in order to see 3c273, identification of the stars in Charts 1a/1b can be made with the naked eye, and those in Chart 2 using binoculars. Telescopes are required for Charts 3 and 4.

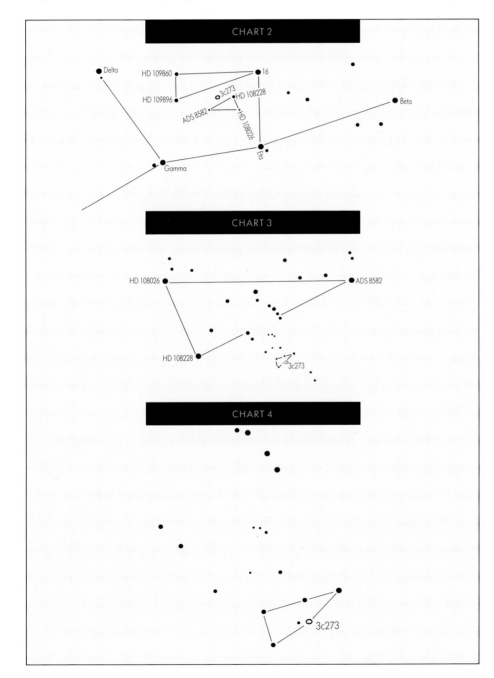

QUASAR 3C273

QUASAR 3c273 DATA

Quasar 3c273 lies at a distance of around 3,000 million light years and has a recessional velocity measured at around 48,000 km (30,000 miles) per second. Its total light output is estimated to be equal to around 300 trillion times that of the Sun, or more than 300 times that of the entire Milky Way Galaxy! The photograph (*above*) shows a luminous jet of material reaching away from the south-west of 3c273. It seems to consist of material ejected from the quasar and is itself a strong emitter of radio waves. Given that the distance-estimates of 3c273 are correct, the jet must be some 300,000 light years long!

MAIN CHART OF NORTHERN SUMMER/ SOUTHERN WINTER STARS

The Summer Triangle, comprising Vega (in Lyra – the Lyre), Deneb (in Cygnus – the Swan) and Altair (in Aquila – the Eagle) dominates the summer night sky for northern hemisphere observers. The distinctive pattern formed by these three stars is unmistakable. Just to the north of Vega is the Head of Draco (the Dragon), a group depicted fully on the main chart of North Circumpolar Stars (see pages 68/69). Further to the south we see the large sprawling constellation of Ophiuchus (the Serpent-Bearer), flanked by Serpens Caput and Serpens Cauda, the Head and Tail of the Serpent. To the south of Ophiuchus is Scorpius, its curved sting prominent and its leading star Antares a true celestial beacon. To the east of Scorpius is Sagittarius, the constellation marking the direction in which lies the centre of our Galaxy.

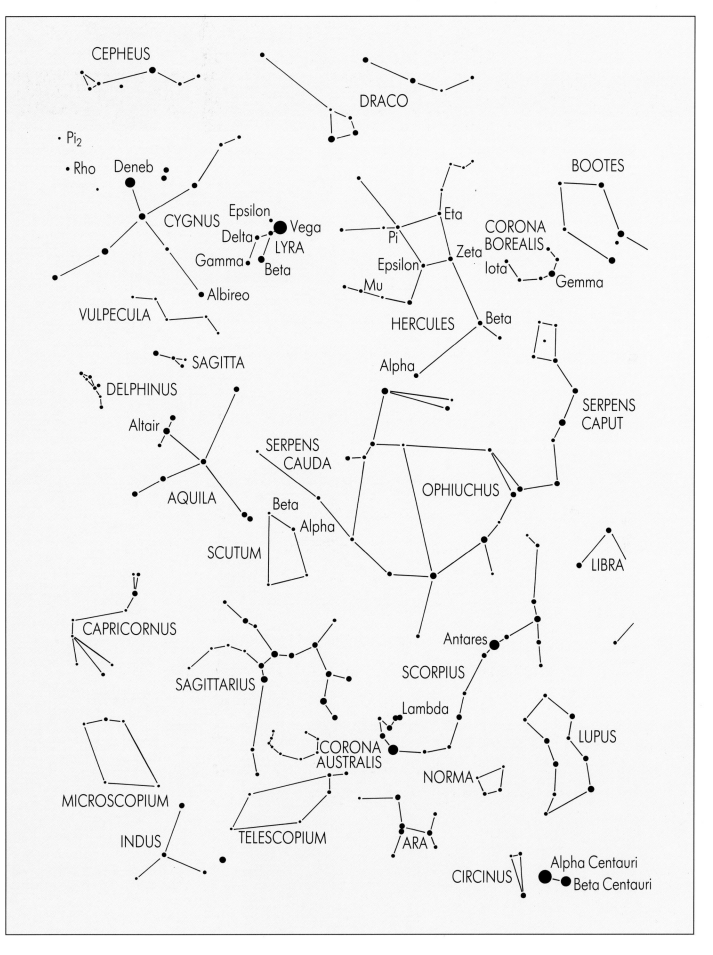

CEPHEUS

DRACO

Pi₂

Rho

Deneb

CYGNUS

Epsilon

Delta

Gamma

Beta

LYRA

Vega

Albireo

VULPECULA

Eta

Pi

Epsilon

Mu

Zeta

HERCULES

Beta

Alpha

BOOTES

CORONA
BOREALIS

Iota

Gemma

SAGITTA

DELPHINUS

Altair

AQUILA

SERPENS
CAUDA

Beta

Alpha

SCUTUM

SERPENS
CAPUT

OPHIUCHUS

LIBRA

CAPRICORNUS

SAGITTARIUS

Antares

SCORPIUS

Lambda

LUPUS

MICROSCOPIUM

INDUS

TELESCOPIUM

CORONA
AUSTRALIS

NORMA

ARA

CIRCINUS

Alpha Centauri

Beta Centauri

97

LYRA

Although Lyra is a small constellation, it contains many objects of interest to the backyard astronomer. Its brightest star – the fifth brightest star in the sky – is brilliant Vega, a magnitude 0·04 object shining from a distance of 27 light years. Its surface temperature is around 9,000°C (16,200°F), roughly twice that of the Sun, its true luminosity being about 58 times that of our star.

Lyra contains the prototype of the Beta Lyrae-type eclipsing binaries. In the Beta Lyrae system, the two stars orbit each other once every 12·908 days. Both components are distorted into ellipsoids, this distortion being a combination of their rapid orbital period and gravitational forces caused by the closeness of the two stars. They are so close that, in fact, their atmospheres intermingle. The magnitudes of Beta Lyrae variables generally vary by no more than a couple of magnitudes, their light curves displaying alternate deep and shallow minima. The magnitude of Beta Lyrae itself fluctuates between a maximum of 3·4 and alternating minima of 3·8 and 4·1. Nightly comparisons of Beta with neighbouring Gamma (magnitude 3·25) over a couple of weeks will enable you to spot its variations (see chart *right*).

Delta Lyrae is a double star easy to resolve using small telescopes or binoculars; its magnitude 4·5 and 5·5 components are separated by a huge 10·5'. The colour contrast between the brighter red star and its fainter white companion is evident, even through binoculars.

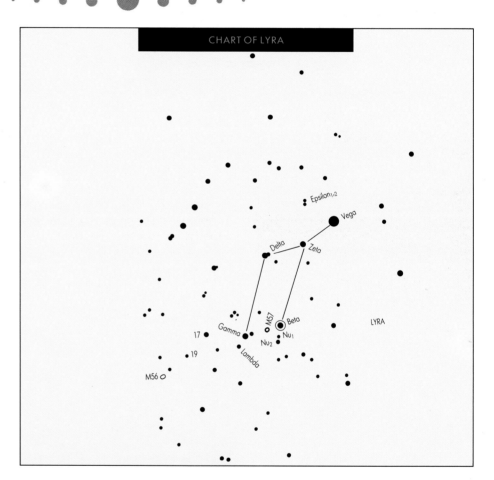

CHART OF LYRA

EPSILON LYRAE AND VEGA

EPSILON LYRAE

One of the most famous objects in Lyra is the Double-Double star Epsilon Lyrae. Located just 1½° to the north-west of Vega, Epsilon appears at first glance to be a single star, although those with really keen eyesight will identify it as a double, as shown in this photograph (*left*). Its 5th and 6th magnitude components lie 208" apart, and any form of optical aid will easily resolve the two stars. Closer inspection with a telescope of at least 100-mm (4-in) aperture will show that each of the two main components is double again, making Epsilon a quadruple star system. The northernmost pair, known as Epsilon 1, has components of magnitudes 5·5 and 6·5 separated by 2·8". The southern pair, Epsilon 2, comprises stars of magnitudes 5·0 and 5·5 located 2·2" apart.

M57 RING NEBULA

Situated at a point just over halfway from Gamma to Beta Lyrae, and a little to the south of a line joining these two stars, the Ring Nebula is perhaps the most famous planetary nebula (see Stellar Evolution, pages 60-63) in the sky. This magnitude 8·8 object, which can be located using the accompanying finder chart (*bottom left*), is beyond the reach of binoculars, although a 75-mm (3-in) telescope will enable you to pick it out.

Instruments of 100-mm (4-in) aperture or more will reveal M57 as a tiny, luminous ring, as seen on this photograph (*below left*). The Ring Nebula, as is the case with other planetary nebulae, consists of material thrown off a star during the latter stages of its evolution. In the case of M57, the ejected material has formed a shell which surrounds its parent star, a dim 15th magnitude object which cannot be seen without a large telescope. As we gaze at M57, we are looking through the nearest wall of the shell of gas. Around the edge is a much deeper, more opaque concentration of material which therefore appears as a ring. The same effect would doubtless be noticed by observers looking at M57 from other positions in space. The neat, well-contained shape of M57 is not typical of planetary nebulae; most of the other 1,000 or so known examples have comparatively irregular forms.

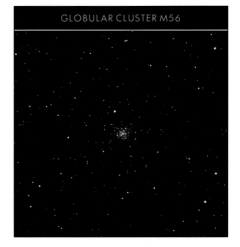

GLOBULAR CLUSTER M56

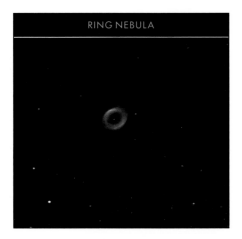

RING NEBULA

GLOBULAR CLUSTER M56

Lying around 5° to the south-east of Gamma Lyrae is the globular cluster M56 (NGC 6779) (see *above* and *below*). This 8th magnitude object lies at a distance of around 46,000 light years and has a diameter of 60 light years. Binoculars will reveal M56 as a small, fuzzy starlike object. Resolution of individual stars around its outer edges is only possible with telescopes of 200-mm (8-in) aperture or more.

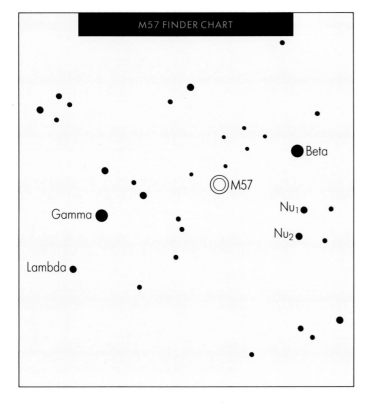

M57 FINDER CHART

Beta
M57
Gamma
Lambda
Nu₁
Nu₂

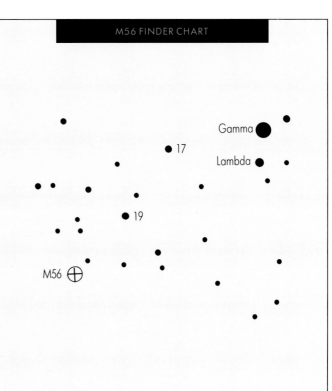

M56 FINDER CHART

Gamma
17
Lambda
19
M56

HERCULES

Hercules is a somewhat scattered group of stars lying to the west of the star Vega (see chart *below*). The brightest star in Hercules is magnitude 2·78 Beta Herculis, also known as Kornephoros. Beta shines from a distance of 105 light years and has a true luminosity some 65 times that of the Sun. Also known as Ras Algethi, Alpha Herculis is an irregular variable, fluctuating between magnitudes 3·1 and around 3·9 over an average period of between 90 and 100 days. It is also a double star, its component of magnitude 5·39 located 4·6" away. The central asterism formed from the stars Zeta, Eta, Epsilon and Pi, can be picked out to the upper left of the photograph which accompanies the section on Corona Borealis (see *opposite*).

GLOBULAR CLUSTER M13

Hercules contains the brightest globular cluster in the northern hemisphere. Also known as the 'Great Hercules Cluster', M13 (NGC 6205) lies at a distance of around 21,000 light years. Its central core measures some 100 light years across, although its outer diameter is roughly twice this. The magnitude of M13 is 5·9, making it dimly visible to the naked eye, under really clear skies. Located roughly a third of the way from Eta to Zeta Herculis, M13 can be easily located with binoculars using the accompanying finder chart, and will be seen nestling between two 7th magnitude stars as shown. Binoculars or finderscopes will show M13 as a fuzzy starlike object, although telescopes of 100-mm (4-in) aperture or more will resolve individual stars around the cluster's edge. Larger instruments, of 300-mm (12-in) aperture and upwards, will provide stunning views of this fine object, with many stars resolved across the entire cluster (see *below right*).

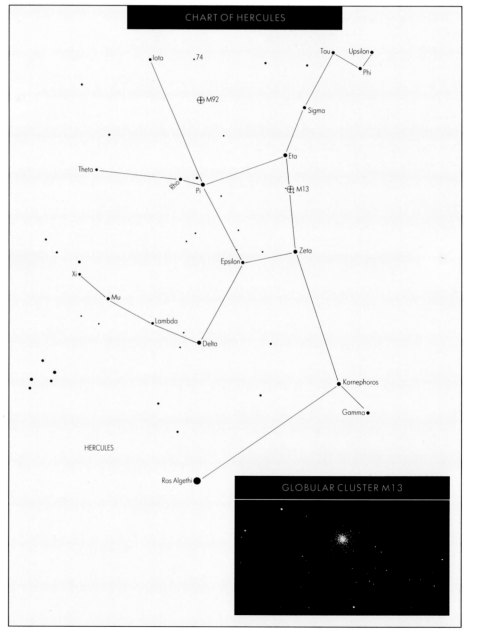

CHART OF HERCULES

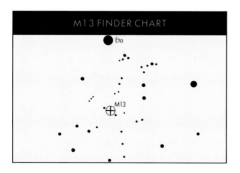

M13 FINDER CHART

GLOBULAR CLUSTER M13

CORONA BOREALIS

Corona Borealis is a conspicuous circlet of stars located to the west of Hercules (see main Northern Summer/Southern Winter chart, pages 96-97 and the photograph shown here *below*). Its leading star is Alpha Coronae Borealis, or Gemma. Shining with a magnitude of 2·23, Gemma lies at a distance of 75 light years and has a true luminosity of around 45 times that of the Sun.

R CORONAE BOREALIS

Located within the circlet of stars forming Corona Borealis (see chart *below*) is the famous variable star R Coronae Borealis. This is the prototype of a class of irregular variable stars which remain at maximum brightness for long periods before suffering sudden reductions in magnitude, followed by a slow recovery to their former glory. R Coronae Borealis has a maximum magnitude of 5·7, and is just visible to the naked eye, under really clear skies. When R Coronae Borealis fades, it can drop as much as nine magnitudes! These reductions take place over periods of a few weeks with minimum brightness lasting several months. However, on several occasions since its behaviour was first noticed in 1795, R Coronae Borealis has remained at or near minimum for a year or more.

The fluctuations exhibited by variables of this type are probably caused by clouds of dark material surrounding the star. This material may be ejected by the star itself and, while it remains around the star, effectively blocks off part of its

CORONA BOREALIS AND PART OF HERCULES

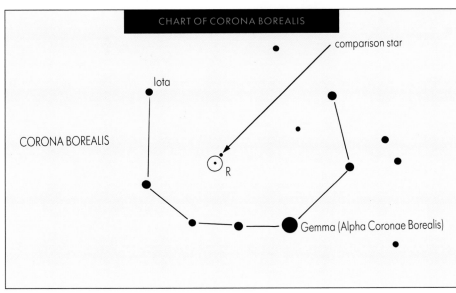

CHART OF CORONA BOREALIS

CORONA BOREALIS

Iota

comparison star

R

Gemma (Alpha Coronae Borealis)

light. Once the material clears, either by being blown off into space or by falling back into the star, the variable regains its former brightness. The unpredictability of stars of this type, means that nightly observation is required in order to detect any sudden dimming. In the case of R Coronae Borealis, there is a handy comparison star of magnitude 7·2 immediately to the north-north-west. Both R Coronae Borealis and the comparison star form a close pair practically midway between Alpha (Gemma) and Iota Coronae Borealis and can be found by using the accompanying finder chart (*left*). If you cannot find the variable, it may be that it is suffering one of its unpredictable falls.

SCORPIUS

The photograph shown here (*below right*) captures most of the constellation Scorpius, the Scorpion, the distinctive curve of the Scorpion's tail visible just to the right of centre at bottom of picture. The five stars towards the lower left of picture belong to the neighbouring constellation of Sagittarius, and can be identified with the accompanying chart (*below left*) which covers roughly the same area of sky.

The brightest star in Scorpius is Antares, seen (*below right*) at the centre of the trio of stars towards the upper right. Antares, whose name is derived from the Greek for 'Rival of Mars', has a strong reddish hue, and certainly vies for prominence when Mars is in the same area of sky. Antares is a super-giant star shining with a magnitude of 0·92, making it the 15th brightest star in the sky. Its diameter is thought to be around 700 times that of the Sun, with a luminosity equal to around 9,000 times that of our star. However, its mass is probably no more than 15 times that of the Sun, due to its extremely low density. The surface temperature of Antares, which shines from a distance of 520 light years, is around 3,000°C (5,400°F). Beta Scorpii is a fine double star for small telescopes, its magnitude 2·6 and 4·9 components being separated by 13·7". Both stars are white; a hint of blue may be detected in the fainter star.

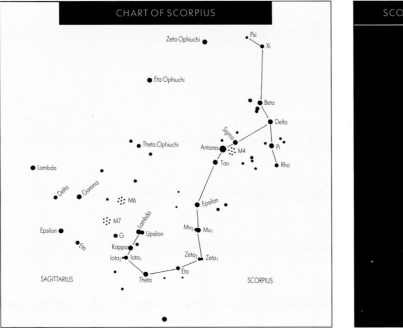

CHART OF SCORPIUS

SCORPIUS AND PART OF SAGITTARIUS

OPEN CLUSTERS M6, NGC 6416 AND H18

The finder chart shown here (*right*) reveals a number of star clusters to be found in Scorpius. Telescopes of 150-mm (6-in) aperture will enable you to see the open cluster H18, an 8th magnitude collection of around 80 stars visible in the same field as M7.

The fine open cluster M6 (NGC 6405) lies 3½° to the north-west of M7. Lying at a distance of over 1,200 light years, its 130 or so stars shine with a combined magnitude of 4·6, making M6 easily visible to the naked eye. As with M7, low-power, wide-field instruments give the best views of M6. The shape of M6 has been likened to a butterfly with open wings, leading to this object being named the 'Butterfly Cluster'. Lying just to the east of M6 is the cluster NGC 6416, a 6th magnitude gathering of around 40 stars shining from a distance of 2,600 light years.

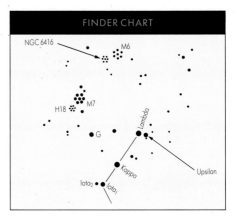

FINDER CHART

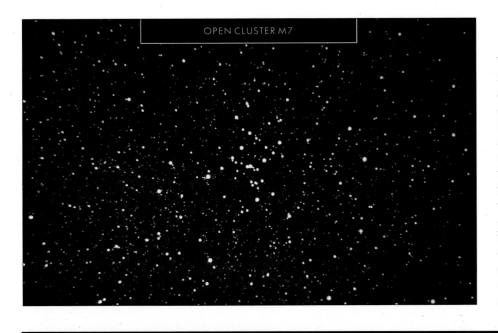

OPEN CLUSTER M7

OPEN CLUSTER M7

The region of sky to the north-east of the Scorpion's tail contains several open clusters, a number of which are observable with only moderate optical aid. Brightest by far of these is 3rd magnitude M7 (NGC 6475), easily visible to the naked eye (see *left*). Low-power, wide-field instruments, such as binoculars or finderscopes, probably give the best views of M7, which is scattered across an area of sky 50' in diameter, nearly twice that of the lunar disc. M7 lies at a distance of around 800 light years and can be picked out using the finder chart (see *opposite bottom*), which shows several other open clusters.

SCUTUM

R SCUTI

The constellation Scutum lies just to the south-west of Aquila and can be easily located using the main chart on pages 96-97. Although small, Scutum contains several objects of interest including R Scuti, a RV Tauri-type variable star (named after the prototype RV Tauri in Taurus). Displaying a ruddy hue, R Scuti is found just 1° south of Beta Scuti (see chart *right*).

RV Tauri variables are very luminous, their light curves generally displaying alternate deep and shallow minima. Discovered by the English astronomer E Pigott in 1795, R Scuti varies between magnitudes 4·8 and 6·0, although every fifth minimum or so sees it drop to 8th magnitude or less.

There are two distinct types of RV Tauri stars, one of which (including RV Tauri itself) undergoes short-term variations common to all members of this class of variable, although these are superimposed on a much longer cycle, which has the effect of raising and lowering the overall magnitudes displayed in the light curve. The other type, whose ranks include R Scuti, do not undergo this superimposed cycle. (See Variable Stars, pages 114-115 for more information.)

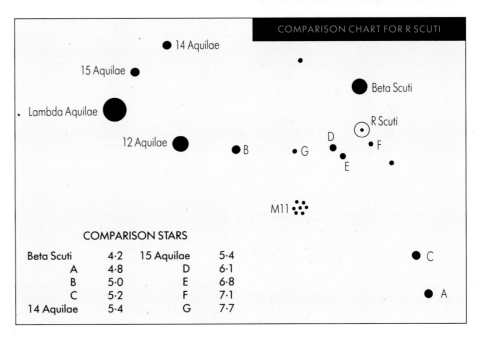

COMPARISON CHART FOR R SCUTI

COMPARISON STARS

Beta Scuti	4·2	15 Aquilae	5·4
A	4·8	D	6·1
B	5·0	E	6·8
C	5·2	F	7·1
14 Aquilae	5·4	G	7·7

OPEN CLUSTER M11

Located in the same general area as R Scuti, is the open cluster M11 (NGC 6705), a gathering of over 500 stars shining from a distance of 5,500 light years. Glowing at 6th magnitude, M11 can just be glimpsed with the naked eye under exceptional viewing conditions. Binoculars bring it out clearly, although magnifications of 20× or more are needed to resolve any individual cluster stars. When seen through binoculars or small telescopes, M11 resembles a globular; larger instruments of 75-mm (3-in) aperture or more start to reveal its distinctive wedge shape. The English astronomer, Admiral William H Smyth, described M11 as resembling '. . . a flight of wild ducks in shape . . .'.

THE MILKY WAY

The Milky Way is visible as a faint pearly band of light stretching right round the sky. Given clear, dark skies, it is easily visible to the unaided eye and any form of optical aid will show that it is made up of many thousands of stars. The Milky Way is actually our view of the Galaxy, looking along the main galactic plane. The pearly glow we see is the combined light from many different stars. Each star, although too dim to register on the unaided eye, contributes to the glow we see.

Our Galaxy is a huge spiral-shaped system comprising in excess of 100,000 million stars. The spiral arms emanate from a central bulge some 20,000 light years in diameter and 10,000 light years thick. Surrounding this bulge is the galactic disc, our Solar System being situated in the disc around 30,000 light years from the galactic nucleus.

THE MILKY WAY IN SCORPIUS AND SAGITTARIUS

This view (see *right*) shows the Milky Way crossing the constellations Scorpius and Sagittarius, the latter containing the centre of the Galaxy. It is in this region of sky that the Milky Way is at its brightest for southern-hemisphere observers. From here it makes its way around the celestial sphere, passing through Norma and Ara, southern Centaurus, Crux and Musca, eastern Carina, Vela, to the east of Puppis and Canis Major, and Monoceros, where it is at its faintest. From here it passes through western Gemini, Auriga, Perseus, Cassiopeia, northern Lacerta, Cygnus, Vulpecula, Sagitta, Aquila, and Scutum before moving into Sagittarius once more.

MILKY WAY IN SCORPIUS AND SAGITTARIUS

THE COALSACK

There are numerous 'gaps' scattered along the length of the Milky Way; one of these is the Coalsack. The prominent cruciform-shape of Crux is seen here (*below*) together with the Coalsack, which at first sight appears to be a dark abyss. Closer inspection reveals it to be a huge patch of dark nebulosity located between us and the stars of the Galaxy.

COALSACK IN CRUX

THE MILKY WAY IN CYGNUS

The brightest stretch of Milky Way for northern-hemisphere observers is that from Cygnus to Aquila. This photograph (*below*) shows the Milky Way passing through Cygnus. The Milky Way contains copious amounts of interstellar gas and dust in the form of nebulae. One of these, the North American Nebula (NGC 7000) can be seen just to the east (left) of the bright star Deneb (Alpha Cygni) near top left of picture.

MILKY WAY IN CYGNUS

THE GALACTIC CENTRE

The huge amounts of interstellar dust which permeate the Galaxy block out radiation at visible wavelengths, the result of which is that we are unable to see down to the heart of the Galaxy. However, other types of radiation can penetrate this matter, including infra-red radiation. These two images (see *below*) were obtained at infra-red wavelengths, the left-hand picture spanning the central 150 light years of the Galaxy. The plane of the Milky Way spans from upper left to lower right, the bright central region being the nucleus of the Galaxy. Many astronomers believe that this region hosts a huge black hole.

The right-hand picture of the Galactic nucleus spans a volume of space 50 light years across and shows a dark red patch to the right of the bright nucleus. This patch is a section of a ring of gas and dust which surrounds the nucleus, and stretches to a radius of around 6 light years.

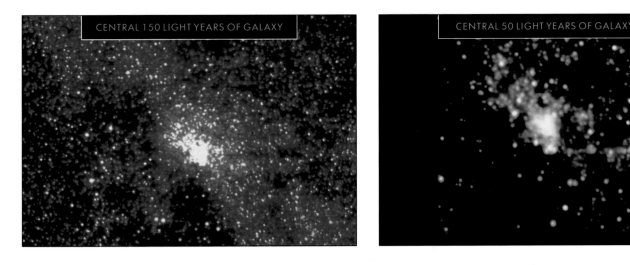

CENTRAL 150 LIGHT YEARS OF GALAXY

CENTRAL 50 LIGHT YEARS OF GALAXY

The landmark of the autumn night sky for northern hemisphere observers is the Square of Pegasus, the quadrilateral of stars forming the main part of the constellation Pegasus. From here, several other groups can be found including Andromeda, which extends from the north-eastern corner of the Square, and Piscis Austrinus, its brightest star Fomalhaut located by extending a line from Scheat through Markab towards the south as shown. To the south-east of Pegasus is the large but dim constellation Cetus, its main claim to fame being the long-period variable star Mira (see pages 110-111 and 114-115). Cetus is located towards the northern end of the constellation Eridanus, depicting the River Eridanus and which is shown to its full extent on the main chart of Northern Winter/Southern Summer stars (see pages 76/77). A celestial aviary can be seen at foot of this chart, comprising Grus (the Crane), Tucana (the Toucan) and Phoenix (the Phoenix), the latter lying just to the north of the bright star Achernar.

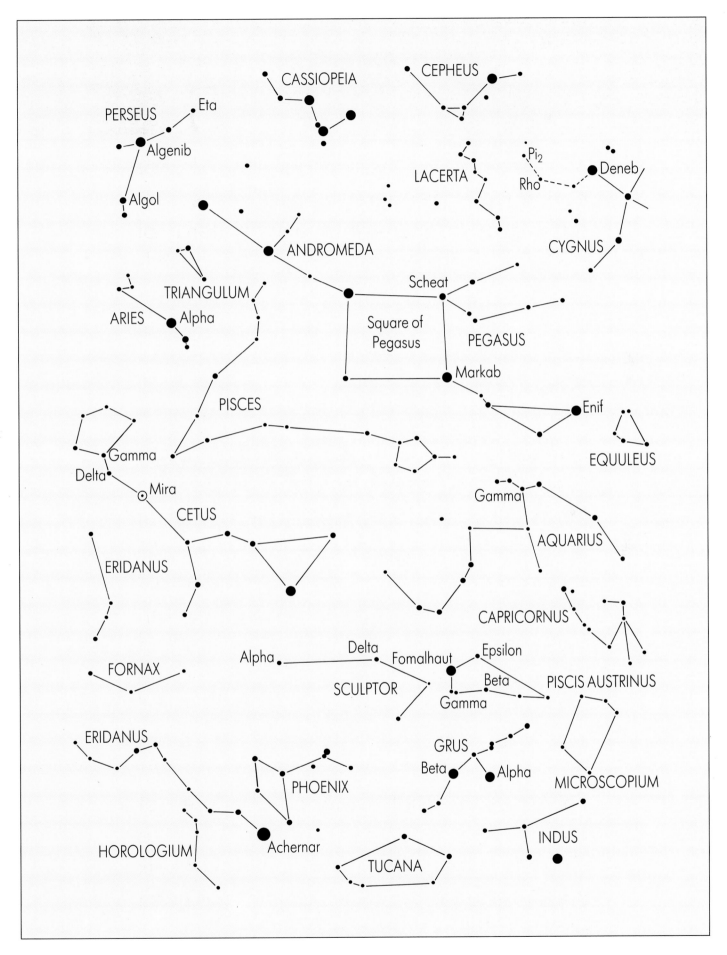

PERSEUS

Located high in the south-east during autumn evenings for observers at mid-northern latitudes, the prominent constellation Perseus contains a wealth of objects of interest to the backyard astronomer. This group, named after the legendary Greek hero, takes the form of a huge inverted 'Y' and is easily located a short way to the south-east of the prominent 'W' of the constellation Cassiopeia.

THE ALPHA PERSEI GROUP

The brightest star in Perseus is magnitude 1·79 Alpha Persei, or Mirfak. This giant white star has a luminosity equal to around 4,000 times that of our Sun and shines from a distance of 570 light years. Perseus contains many rich starfields, one of particular note being the Alpha Persei Group, a stellar association centred on Alpha Persei which contains over 100 stars. Stellar associations are groups of stars that formed together but whose combined gravitational attraction was insufficient to maintain their proximity to each other. The speeds of travel of the member stars are so great that, unlike open clusters, they simply drift apart as they move collectively through space. It is thought that the stars in the Alpha Persei Group formed around 4 million years ago but have been moving away from each other ever since.

ALGOL

Algol (Beta Persei) is one of the most famous variable stars in the entire sky. Its name is derived from the Arabic 'Al Ra's al Ghul' meaning 'The Demon's Head' and, according to legend, it depicts the severed Head of Medusa the Gorgon held by Perseus. It seems that Algol's variability was known to Arabic astronomers around the 10th century, although it was the English astronomer John Goodricke who made the first accurate measurements of its period of variability. It was also Goodricke who first suggested that Algol was a binary and that the variations in brightness were due to the occasional eclipse of a brighter star by a fainter companion.

Algol is the prototype eclipsing binary and, at a distance of around 100 light years, is the closest object of its kind. As Goodricke stated, it is indeed a system of two stars in orbit around each other, with one star considerably fainter than its companion. The plane of their orbit is almost exactly lined up with our position in space and the fainter star regularly passes in front of, or eclipses, its brighter companion. When this occurs, the overall light output from Algol decreases

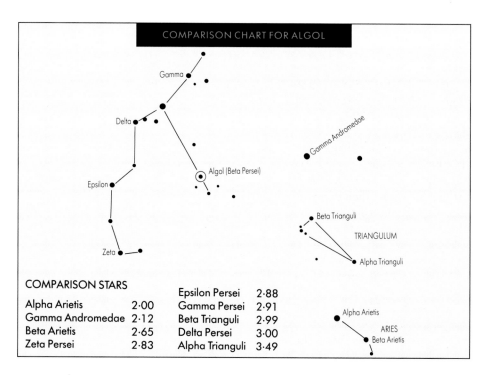

COMPARISON CHART FOR ALGOL

COMPARISON STARS			
Alpha Arietis	2·00	Epsilon Persei	2·88
Gamma Andromedae	2·12	Gamma Persei	2·91
Beta Arietis	2·65	Beta Trianguli	2·99
Zeta Persei	2·83	Delta Persei	3·00
		Alpha Trianguli	3·49

from magnitude 2·1 to 3·4 before climbing again (see Variable Stars, pages 114-115). The entire sequence takes around 10 hours with a well-determined period between successive minima of 2·86739 days (2d 20h 48m 56s).

To observe Algol's variations, compare its brightness during an eclipse sequence with those of the stars shown on the accompanying chart (*above*). This chart shows the positions of stars whose magnitudes lie roughly within Algol's range of magnitudes. By making a series of random checks on Algol over several nights, you should detect its fluctuations. Look out for predictions of minima, which appear in numerous astronomical publications.

THE SWORD HANDLE DOUBLE CLUSTER

Located near the border between Perseus and Cassiopeia, roughly midway between the main stars of each group (see finder chart *below right*), the two open clusters NGC 869 and NGC 884 are visible to the naked eye on clear nights. Collectively known as the Sword Handle Double Cluster, attention was drawn to them as long ago as 150 BC by the Greek astronomer Hipparchus.

SWORD HANDLE DOUBLE CLUSTER

NGC 869 shines from a distance of around 7,000 light years while NGC 884 is located some 8,150 light years away. Their overall magnitudes are 4·4 (NGC 869) and 4·7 (NGC 884). Both clusters have diameters of around 70 light years. The stars within these clusters are very hot and luminous. Their combined mass is equal to some 5,000 times that of the Sun, although their true luminosity is 200,000 greater than that of our star. Binoculars will show the clusters quite well, and small telescopes will reveal the splendour of these two beautiful objects. As they lie half a degree apart, a low-power eyepiece and wide field of view are required to view both clusters at once. (See photograph *above*.)

The Sword Handle Double Cluster lies in a rich section of the Milky Way and time spent sweeping this area with binoculars or a wide-field telescope is well rewarded. Follow the example of the English observer Rev. Thomas William Webb when he points out that 'Night after night the telescope might be employed in sweeping over its magnificent crowds of stars . . .'

OPEN CLUSTER M34

Discovered by Charles Messier in August 1764, the open cluster M34 (NGC 1039) is an easy object for binoculars or a small telescope. It can be tracked down by star hopping from Algol as shown on the accompanying finder chart (see *below*). If you have really keen eyesight, and the sky is very dark and clear, you may glimpse this magnitude 5·5 object with the naked eye.

M34 lies at a distance of around 1,500 light years and contains around 80 stars within a volume of space 18 light years across. Binoculars with magnifications of up to 10× will show M34 as a

OPEN CLUSTER M34

distinct nebulous patch, magnifications above this enabling individual stars to be resolved. When used with low-power eyepieces, small telescopes will show M34 as a large scattered group of stars (see photograph *above*).

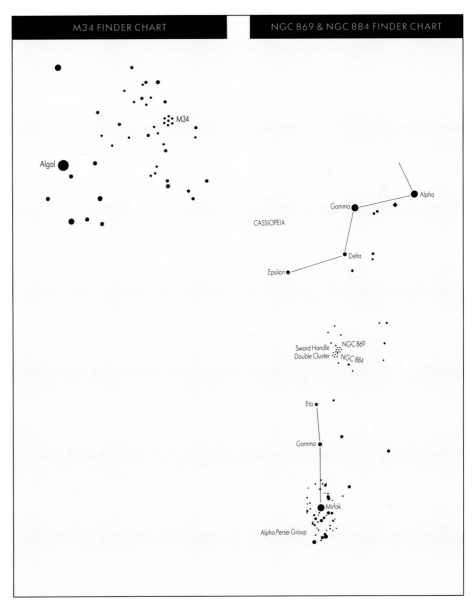

M34 FINDER CHART

NGC 869 & NGC 884 FINDER CHART

CETUS

The large, sprawling and somewhat inconspicuous constellation of Cetus can be seen in this chart (*below right*). Compare this chart with the photograph shown on the opposite page. Mira, the reddish star near top right corner of picture, appears near maximum brightness. Delta Ceti is the blue-white star to the upper left of Mira. Some of the other stars in Cetus can also be seen in this photograph, which also includes several members of the neighbouring constellation of Eridanus. The double star Gamma Ceti is just resolvable in 100-mm (4-in) telescopes under good seeing conditions, its magnitude 3·6 and 6·2 white and yellowish components separated by 2·7". Another double just resolvable in 100-mm (4-in) telescopes is Nu Ceti which has magnitude 4·9 and 9·5 components separated by 8·1". Their colours are yellowish and white.

SPIRAL GALAXY M77

This 9th magnitude face-on compact spiral galaxy is located 1° to the southeast of Delta Ceti, as shown on the accompanying finder chart (*below*). Located at a distance of 60 million light years, M77 (NGC 1068) has an overall diameter of around 100,000 light years. Under dark, clear skies, telescopes of 75-mm (3-in) aperture will reveal M77 as a tiny fuzzy ball, instruments of 150 mm (6 in) bringing out the bright, central regions together with the much fainter surrounding disc (see *below right*). M77 is a 'Seyfert' galaxy, a class of galaxy first studied by the American astronomer Carl Keenan Seyfert in 1943. These galaxies have bright, compact, almost starlike nuclei and are strong sources of infra-red and ultra-violet emissions.

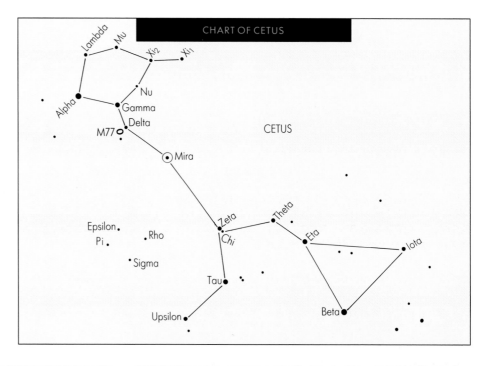

CHART OF CETUS

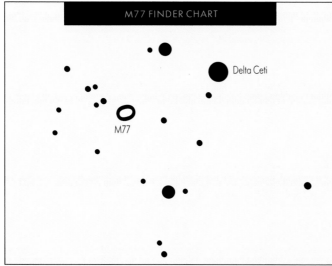

M77 FINDER CHART

COMPACT SPIRAL GALAXY M77

Mira holds the distinction of being the first variable star to be discovered. On 13 August, 1596, the Dutch astronomer David Fabricus observed the star, mistaking it for a nova. It was next spotted in 1603 when the German astronomer Johann Bayer catalogued it as a 4th magnitude star and included it as Omicron Ceti in his famous star atlas, the *Uranometria*. Not long after this Omicron Ceti disappeared from view, only to reappear almost a year later. Subsequent observation finally revealed its true nature. Because Omicron Ceti was the first star known to vary in brightness, it was regarded as being highly unusual. The German astronomer Johann Hevelius named the star Mira, meaning 'wonderful'. Since the discovery that Mira was variable in 1638 every maximum has been observed.

Mira is the prototype long-period variable star, this class of variable being the largest known (see Variable Stars, pages 114-115). Around 4,000 Mira-type stars have now been catalogued, all of which are red giants. Their amplitudes average out at around 5 or 6 magnitudes, although some are known to vary by as much as 9 or 10 magnitudes. The variations are by no means regular, and substantial differences between successive cycles of variability occur. Also, their periods can be anything from less than 100 days to 700 days or more, although successive periods can differ markedly. The period of Mira, although averaging out at 331 days, has been known to vary dramatically, durations as short as 304 days and as long as 355 days having been recorded.

Mira is the brightest of the long-period variables and only moderate optical aid is required to follow its complete cycle. It is easily located roughly two-fifths of the way from Delta to Zeta Ceti, as shown on the main chart (see *opposite page*). Delta Ceti and Mira, the two stars in the top right of picture, can be seen in this photograph (*below left*). Cetus itself is best placed for observation during November when it will be visible above the southern horizon during mid- to late-evenings. Observers in the southern hemisphere, of course, will see Cetus above the northern horizon. Continued observation of Mira over a period of several weeks will reveal its changes in magnitude against the comparison stars included on this chart (*below*). As is the case with any variable star which remains at or near minimum brightness for extended periods, Mira may well be out of view when you look for it. If this is the case, then keep an eye open for it. Mira will eventually reappear, following which magnitude estimations should be made on a weekly basis.

PART OF CETUS, SHOWING MIRA AND DELTA CETI

MIRA (OMICRON CETI)
Long-period pulsating variable

Normal maximum brightness:	3rd or 4th magnitude
(reached almost 1st magnitude in 1779)	
Normal minimum brightness:	9th or 10th magnitude
Period:	331 days on average

COMPARISON STARS

Beta Ceti	2·00	70 Ceti	5·41
Alpha Ceti	2·52	G	6·00
Eta Ceti	3·44	71 Ceti	6·40
Mu Ceti	4·26	F	6·49
Xi$_2$ Ceti	4·27	A	7·19
Xi$_1$ Ceti	4·36	B	8·00
Lambda Ceti	4·70	C	8·60
Nu Ceti	4·87	D	8·80
75 Ceti	5·34	E	9·20

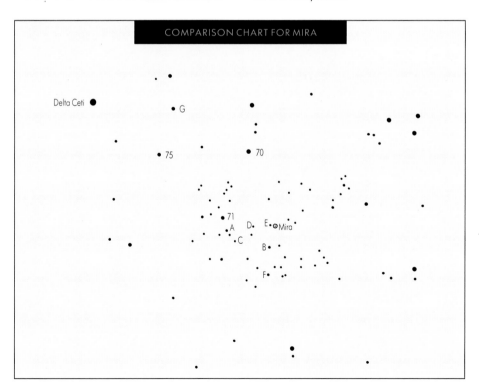

COMPARISON CHART FOR MIRA

Delta Ceti

G

75   70

71   D   E Mira
A   C   B
F

AQUARIUS

This photograph (*below left*) shows the bright star Fomalhaut, visible near bottom margin. Fomalhaut is the leading star of the southern-hemisphere constellation Piscis Austrinus and the 18th brightest star in the sky. Also known as the 'Solitary One', Fomalhaut is located in a somewhat barren region of sky. It shines from a distance of 23 light years and has a true luminosity some 14 times that of our Sun.

Just to the north-west (upper right) of Fomalhaut is the star Epsilon Piscis Austrini, a star used on the finder chart for the Helix Nebula (see *bottom right*). A little way to the north (above) of Epsilon is a tiny triangle of faint stars comprising 66 (the reddish top star in the triangle), 68 (to the left of 66) and 59 Aquarii, the latter two being quite faint. These three stars also appear in the Helix Nebula chart.

FOMALHAUT AND STARS OF AQUARIUS

HELIX NEBULA

HELIX NEBULA

This chart (*right*) shows how to locate the Helix Nebula (NGC 7293) from the bright star Fomalhaut. The Helix Nebula is located immediately to the right (west) of the faint star 59 Aquarii.

The Helix is the largest and closest known planetary nebula. Estimates put its distance at around 500 light years, although opinions are divided as to its exact distance. Like other planetary nebulae, the Helix Nebula is a shell of gas thrown off by a star which then collapses to form a white dwarf (see the photograph *above right*). The gas within the Helix (and other planetary nebulae) shines as a result of the energy emitted by the nebula's central stars, the newly-exposed surfaces of which can have temperatures of up to 100,000°C (180,000°F) or more.

The Helix Nebula is an elusive object. Shining with a magnitude of around 6½,

its overall surface brightness is quite low and it is virtually unobservable except through either a pair of binoculars or a wide-field telescope equipped with a low-power eyepiece. It is a southern hemisphere object and, except under ideal conditions, those at or near mid-northern latitudes will have their work cut out to find it as it lies only a few degrees or so above the southern horizon. Southern-hemisphere observers, and those in the equatorial regions, will be able to find the Helix more easily.

In order to find it, clear, dark, moonless skies are essential. Northern-hemisphere observers should pick a time when Aquarius is at or near its highest elevation. Binoculars show the Helix as a nebulous, circular patch of light, telescopes of 300 mm (12 in) or more bringing out the overall shape of the nebula. The maximum effective telescopic magnification for the Helix is equivalent to 6× per cm (15× per in) of aperture.

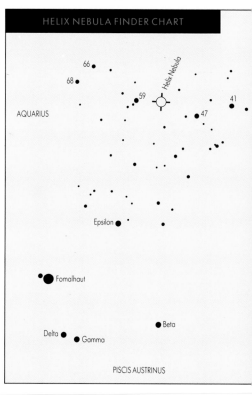

HELIX NEBULA FINDER CHART

AQUARIUS

66
68
59
Helix Nebula
41
47

Epsilon

Fomalhaut

Beta

Delta Gamma

PISCIS AUSTRINUS

SCULPTOR

SPIRAL GALAXY NGC 253

This chart (*below*) shows the constellation of Sculptor, a faint group located immediately to the east of Fomalhaut. The brightest star in Sculptor is magnitude 4·37 Beta, Gamma being almost as bright at magnitude 4·41.

In spite of its obscurity, Sculptor hosts several deep-sky objects of interest, notably the 7th magnitude spiral galaxy NGC 253, found to the north of Alpha Sculptoris. NGC 253 is roughly half the diameter of our own Galaxy and is orientated practically edge-on to us, appearing as an elongated, almost needle-like patch of light when viewed through binoculars or small telescopes (see photograph *right*). It is the leading member of the Sculptor group of galaxies, a collection of island universes situated at a distance of around 8 million light years, and ranks second only to the Andromeda Spiral and the Magellanic Clouds in terms of observability. Like the Helix Nebula, NGC 253 remains low over the southern horizon for northern-hemisphere observers, although its high surface brightness makes it somewhat easier to observe. Telescopes of 250-mm (10-in) aperture or more will start to bring out the mottled surface of NGC 253, caused by huge clouds of

SPIRAL GALAXY NGC 253

obscuring dust scattered across the galaxy.

Lying just to the south and slightly to the east of NGC 253 is the globular cluster NGC 288, an 8th magnitude object discernible as a faint patch of light through binoculars. Located around 2° apart, both NGC 253 and NGC 288 can be seen in the same binocular field

of view. The Mira-type variable star S Sculptoris can be found by extending an imaginary line from Delta through Zeta Sculptoris as shown on the chart (see *below*). S Sculptoris varies between magnitudes 5·5 and 13·6 over a period of 366 days and can be followed through its entire cycle with telescopes of 200-mm (8-in) aperture.

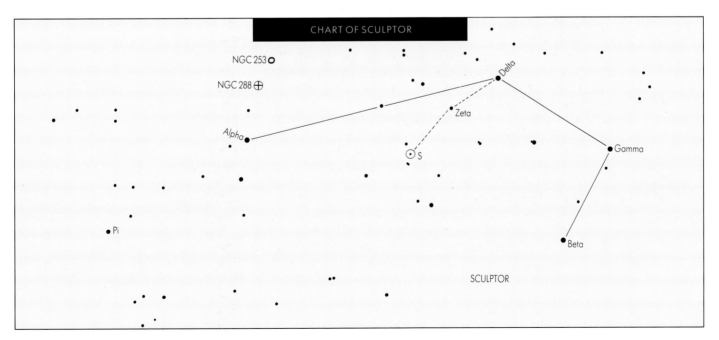

CHART OF SCULPTOR

NGC 253

NGC 288

Delta

Zeta

Alpha

S

Gamma

Pi

Beta

SCULPTOR

VARIABLE STARS

Most stars shine steadily, their energy output remaining fairly constant even over periods of millions of years. Yet there are many stars whose luminosity varies. These stars are known as variables and, depending on the type, their changes take place over periods ranging from less than an hour to hundreds of days. The types of variable stars, together with specific examples, are detailed below.

TYPES OF VARIABLES

There are many types of variable star, each of which can be placed into one of two broad categories. Extrinsic variables are not true variable stars at all, their light output varying due to the influence of another object! Intrinsic variables fluctuate because of changes taking place within the star itself.

ECLIPSING BINARIES

The eclipsing binaries are extrinsic variables, their changes in light output being due to the fact that these are binary star systems. There are several classes of eclipsing binary, the best known of which are the Algol type (see below and pages 108-109) and the Beta Lyrae systems (see pages 98-99).

SHORT-PERIOD VARIABLES

The most famous short-period variables are undoubtedly the Cepheids (see below). They are named after Delta Cephei, the first variable of its type to be discovered (see pages 72-73). Some short-period variables fluctuate over much shorter time spans. The rare and highly luminous Beta Canis Majoris stars undergo only sight changes in magnitude over periods which are measured in hours. Beta Canis Majoris itself hovers around magnitude 2·0, changing by only around 0·03 of a magnitude. Needless to say, these changes are very difficult to detect and their observation is generally beyond the scope of amateurs.

LONG-PERIOD VARIABLES

The variables described above all undergo changes in luminosity over relatively short periods. Different to these are the long-period variables which fluctuate over periods which can extend to a year or more. The most famous long-period variable, and the prototype of the class, is Mira, or Omicron Ceti (see below and pages 110-111).

SEMI-REGULAR VARIABLES

Most semi-regular variables are red giant stars. Their periods of variability, which are not well defined to start with, are subject to variations. These variations are unpredictable as in the case of R Ursae Minoris. This star is similar in some respects to variables in the long-period class, although there are marked differences in the light curves from one period of variability to the next. The magnitude of R Ursae Minoris ranges between 8·5 and 11·5 over a period of 326 days.

IRREGULAR VARIABLES

Irregular variables are giant stars with no regular, well-defined period of variability. Their ranks include the RV Tauri stars, the brightest example of which is R Scuti (see pages 102-103).

ERUPTIVE VARIABLES

This type of variable embraces mainly novae and stars with similar characteristics. Novae are hot dwarf stars which undergo sometimes huge increases in magnitude over very short timespans, usually measured in days. All novae are members of close binary systems in which material is being dragged from a large star onto the surface of a smaller companion, usually a white dwarf. As the depth of the material deposited increases, the temperature and pressure at its base builds up until a critical point is reached where nuclear reactions are sparked off. These reactions produce a nova, which we see as a temporary increase in the brightness of the star as the material is ejected into space.

Over a dozen bright novae have been recorded this century, a fine example being Nova Puppis, or CP Puppis, seen in 1942. This object rose from obscurity to a maximum magnitude of 0·4 during early-November 1942 after which it slowly faded, reaching the threshold of naked-eye visibility by the end of November. It is not known with certainty just what the pre-outburst magnitude of CP Puppis was, although it is believed to have been less than 17th magnitude.

Similar in many ways are the recurrent novae, stars which have been observed to undergo more than one nova-like outburst. A famous example is T Coronae Borealis. Corona Borealis also contains the famous irregular variable R Coronae Borealis, a star which suffers unpredictable *reductions* in brightness. R Coronae Borealis is the prototype of its class and is described on pages 100-101.

Considerably more spectacular than novae are the much rarer supernovae. Described more fully under 'Stellar Evolution', supernovae are massive stars which, having reached the end of their careers, explode violently. During the explosion much of the star is violently ejected into surrounding space, the star itself being virtually destroyed.

LIGHT CURVES

The changes in magnitude of a variable star are plotted by use of a light curve. This is a kind of graph which shows how the light from the star varies between maximum and minimum magnitude over time. The basic layout of a light curve is with magnitude plotted on the vertical axis and time along the horizontal axis. The units used on each axis are dictated by the type of variable. Those whose variations are only very slight, such as the Beta Canis Majoris stars, would have brightness plotted in tenths, or even hundredths, of a magnitude, unlike those of long-period variables where units of whole magnitudes are used. Similarly, variables whose brightness changes quickly would be recorded on a light curve where the time axis is plotted in minutes and hours rather than days, months or even years. The difference between greatest and least brilliance of a variable star is known as the amplitude, while the time taken for a complete cycle is referred to as the period.

ALGOL LIGHT CURVE

The light curve of Algol (*below*) displays alternate deep and shallow minima, arising from the fact that Algol is an eclipsing binary. The Algol system comprises two stars, one of which is much brighter than the other. Their orbital plane is more or less aligned with our position in space resulting in the regular eclipsing of the brighter star by its faint companion and a corresponding reduction in Algol's overall brightness (1 and 3). A secondary minimum occurs when the fainter star passes behind the bright component, although the reduction in magnitude is only very slight, as indicated by the light curve (2).

DELTA CEPHEI LIGHT CURVE

The periods of Cepheids are short and of between 1 and 55 days duration. All 500 or so known Cepheids are very luminous yellow or white giant stars. Their light curves, typified by that of Delta Cephei itself (see *below*), display a slow fall to minimum brightness followed by a much smoother climb to maximum.

MIRA LIGHT CURVE

Mira (Omicron Ceti) is the prototype long-period variable and the first variable star to be discovered. As with all long-period variables, Mira's variations in brilliance are not regular and occur over irregular periods; Mira's amplitude averages 6 magnitudes and its period averages 331 days (see *below*).

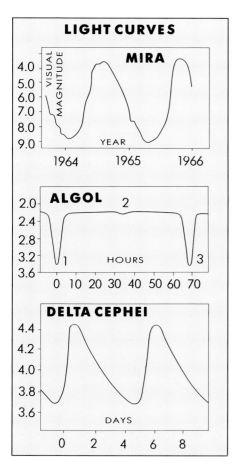

LIGHT CURVES

MIRA

ALGOL

DELTA CEPHEI

OBSERVATION

There are huge numbers of variable stars, many of which can be observed with the naked eye. Binoculars increase the number observable, while even a small telescope will enable hundreds to be observed. Some types of variable are more suited to amateur observation than others. Long-period variables, with their huge ranges in magnitude, make fascinating targets for the backyard astronomer. Also, because their variations can be predicted in advance, eclipsing binaries are popular amongst amateurs, as are the reliable Cepheids, many of which remain at naked-eye visibility throughout their cycle of variability. Observing variable stars involves comparing their changing magnitudes with those of nearby stars of known magnitude. The positions of variable stars, together with the locations and magnitudes of nearby comparison stars, are plotted on special comparison charts, a number of which are included with the star charts in this book.

ESTIMATING MAGNITUDES

There are two different methods of estimating magnitudes, one of which is the Fractional Method in which two comparison stars are used for each magnitude estimation. This seems to be the most popular among variable-star observers. With the Fractional Method, one of the comparison stars should be brighter than the variable at the time of observation, the other fainter. The difference in brightness between the two is scaled mentally into a number of parts. The variable is then placed into its appropriate position in this scale. For example, let's say we have two comparison stars of magnitudes 6·5 (star A) and 7·0 (star B). This difference in magnitude can be divided into 5 parts, each part being equal to 0·1 magnitude. Given that the variable is 2 parts of the way from star A to star B, we would enter its magnitude at time of observation as 6·5 + (2 × 0·1) = 6·7. The result can be checked by using other pairs of comparison stars to see if similar values are derived.

The alternative is the Pogson Step Method whereby the magnitude of the variable is directly compared to that of as many comparison stars as possible. The observer must be able to judge star brightnesses to an accuracy of 0·1 magnitude, a value classed as one 'step' in the Pogson Step Method.

As an example, let's say that the variable is being compared with two comparison stars. You may estimate the variable to be one step brighter than comparison star C and two steps fainter than comparison star D. Looking up the magnitudes of the comparison stars may give values of 7·3 (star C) and 7·0 (star D). This would place the variable at magnitude 7·2. However, wherever possible, double check your results with other comparison stars, some brighter and some fainter than the variable. Ideally, you should reach the same value. If the follow-up estimates are more than 0·1 magnitude out, then something has gone wrong!

GALAXIES

T he Milky Way Galaxy is a huge collection of stars, gas and dust and is just one of countless millions of galaxies scattered throughout space. The diameters of these galaxies range from several thousand to 100,000 light years or more. A very small number are visible without optical aid. Observation has shown that wherever we point our telescopes in the sky, galaxies are found right out to the edge of our limit of vision.

THE HUBBLE CLASSIFICATION

To classify galaxies, astronomers use the system devised by the American astronomer Edwin Hubble. The Hubble system classifies galaxies according to how they appear from Earth, the different types being presented in the now-famous 'tuning fork' diagram. This contains the three main types of galaxy; elliptical (E), spiral (S) and barred spiral (SB). Elliptical galaxies have no spiral arms and are symmetrical in appearance, while spiral galaxies contain a rounded central nucleus from which spiral arms emanate. Barred spiral galaxies are different from the 'normal' spirals in that they have a straight bar which crosses the central bulge, the spiral arms emerging from the ends of this bar.

All three types of galaxy are further classified. Ellipticals are graded from E0 (almost spherical) to E7 (highly elongated), while the spirals are graded according to both the size of their central bulges and to how tightly wound are their spiral arms. An Sa spiral (or SBa in the case of a barred spiral) would have a large and dominant nuclear bulge with tightly wound spiral arms. In an Sc (or SBc) system the roles are reversed, with wide spiral arms and a much smaller nucleus. Sb or SBb galaxies have arms and nuclear regions which are more or less equal in prominence.

Often included on the Hubble diagram (although not originally by Hubble himself) are the transitional (S0) and Irregular (I) galaxies. S0 galaxies have a central region which resembles an elliptical. Around this is a flattened disc, although the disc displays no evidence of spiral structure. Irregular galaxies, as their name suggests, have no symmetrical structure whatsoever and resemble nothing more than loose and randomly distributed collections of stars.

PINWHEEL GALAXY M33

Discovered by the French astronomer Charles Messier in 1764, the Triangulum Spiral Galaxy (M33), also known as the Pinwheel Galaxy, is a member of the Local Group, a collection of galaxies of which our own Milky Way is also a member. Lying at a distance of 2·4 million light years, this system is an Sc spiral, its tiny nucleus and huge, sweeping spiral arms being typical of this class of galaxy.

Spiral galaxies are generally comprised of a mixture of stars and nebulosity. Their nuclear regions contain old stars with little in the way of interstellar gas and dust, while the spiral arms are home to much younger stars with gas and dust clouds. The spiral arms of M33 are irregularly shaped and contain many clumps of nebulosity. The brightest of these is NGC 604, seen as a small bright patch towards the lower left (north-east) edge of the galaxy (see photograph *above right*). Measuring

SPIRAL GALAXY M33

ELLIPTICAL GALAXY M87

1,000 light years across, NGC 604 is one of the largest emission nebulae known.

ELLIPTICAL GALAXY M87

Elliptical galaxies are both the largest and smallest type of galaxy in the Universe. The largest galaxies known are ellipticals, while huge numbers of dwarf elliptical systems have been catalogued. Elliptical galaxies comprise older stars, typical of those found in galactic nuclei and globular clusters. Unlike the outer regions of spiral galaxies, there is little if any interstellar material present in elliptical galaxies, and star formation ceased long ago in these systems. M87 is an E1 type galaxy, and is in fact the most massive galaxy known (see photograph *left*). Situated at the heart of the Virgo Cluster of galaxies, M87 is the cluster's dominant member although optically, as is the case with all ellipticals, it is not very notable.

ELLIPTICAL GALAXY NGC 5128

All galaxies give out radio emissions, the main source of these being the regions of interstellar gas present in them. Indeed, by charting galaxies at radio wavelengths, astronomers are able to map out the positions and extent of the hydrogen clouds found within them. Some galaxies give out huge amounts of radio emission, so much so that they are known as radio galaxies. Their radio output far exceeds that of normal galaxies. Also, unlike normal galaxies, the emissions from radio galaxies come from electrons moving at high speeds rather than from clouds of gas. Nearly all radio galaxies are elliptical, such as NGC 5128 seen here (*right*), which lies some 10 to 15 million light years away and is the closest and most intensively studied radio galaxy in the sky.

The strong radio emissions given out by NGC 5128 emanate from two colossal radio 'lobes' which reach out to a distance of around 1·5 million light years to either side of the galaxy. These lobes, which are invisible at optical wavelengths, consist of material which is moving at very high speeds and which appears to have been thrown out from the central regions. Optically, NGC 5128 is a strange sight, and its exact classification is uncertain. It seems to have a central elliptical- (or, perhaps, S0-) type component which appears to be surrounded by a disc consisting of stars, gas, and dust. This disc is seen as a dark band crossing the much brighter central region. It has been suggested that NGC 5128 is the product of a collision between a spiral and an elliptical galaxy, an idea that seems to be gaining in favour.

ELLIPTICAL GALAXY NGC 5128

BARRED SPIRAL GALAXY NGC 1365

BARRED SPIRAL GALAXY NGC 1365

NGC 1365 is the largest member of the Fornax cluster of galaxies (see *right*) and is regarded as the finest example of a barred spiral galaxy (see *above*). Its central bar has a length of 45,000 light years. From the ends of this bar, long spiral arms curve away. These arms contain bright, luminous regions while the dark lanes, seen both on the central bar and on the insides of the spiral arms, are regions of dark dust which hide the light from stars beyond them. NGC 1365 has an overall diameter exceeding 150,000 light years, making it one of the largest galaxies known.

THE FORNAX CLUSTER OF GALAXIES

Most galaxies are members of groups or clusters which can contain anything from a dozen or so, to over a thousand individual galaxies. Our own galaxy, the Milky Way, is a member of the Local Group, which is a gathering of two dozen or so galaxies.

The Fornax Cluster shown here (*below*), lies at a distance of 60 million light years. Most of the cluster lies within the constellation Fornax, although some members 'spill over' into neighbouring Eridanus. The Fornax Cluster contains a large number of dwarf systems together with around 20 bright galaxies. These include examples of ellipticals, spirals and barred spirals, one of which is NGC 1365, seen here on the bottom margin a little way to the right of centre.

Much larger than the groups and clusters are the superclusters of which many examples have been found. The Local Supercluster contains many individual clusters, including the Local Group located near its edge, and has a diameter of around 100 million light years.

FORNAX CLUSTER

MAIN CHART OF SOUTH CIRCUMPOLAR STARS

Unlike the region around the north celestial pole, that surrounding its southern counterpart contains several faint and somewhat obscure groups. The south celestial pole itself lies in the constellation Octans (the Octant), its position marked (or nearly so) by the faint 5th magnitude star Sigma Octantis (see pages 120/121). The south celestial pole lies on a line between Crux and Achernar, the brighest star in Eridanus, roughly midway between the two. This region of sky, although containing mainly faint stars and constellations, is circled by the bright stars Alpha and Beta Centauri (the Centaur), Canopus (in Carina — the Keel) and Achernar. Also worthy of mention here are the two nearby galaxies, the Large and Small Magellanic Clouds. These objects are clearly visible to the naked eye when the sky is reasonably dark and clear, and are covered in more detail on pages 120-121 and 122-123.

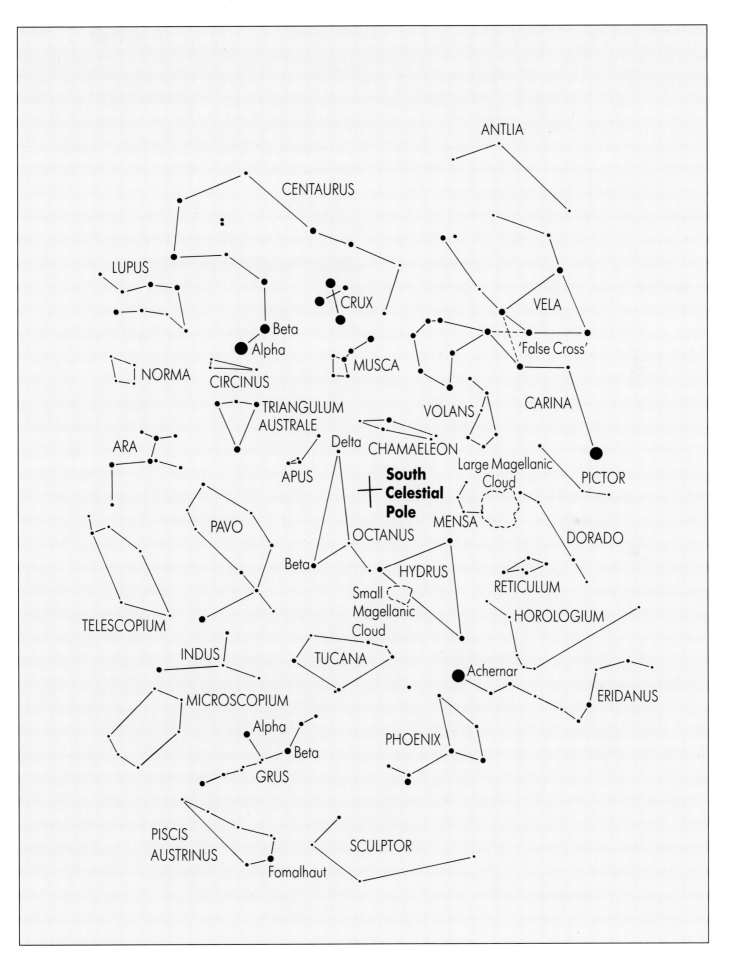

ANTLIA

CENTAURUS

LUPUS

CRUX

Beta

Alpha

NORMA

CIRCINUS

MUSCA

VELA

'False Cross'

CARINA

TRIANGULUM
AUSTRALE

VOLANS

ARA

Delta

CHAMAELEON

APUS

South
Celestial
Pole

Large Magellanic
Cloud

PICTOR

PAVO

OCTANUS

MENSA

DORADO

Beta

HYDRUS

RETICULUM

TELESCOPIUM

Small
Magellanic
Cloud

HOROLOGIUM

INDUS

TUCANA

MICROSCOPIUM

Achernar

ERIDANUS

Alpha

Beta

PHOENIX

GRUS

PISCIS
AUSTRINUS

SCULPTOR

Fomalhaut

Unlike the region of sky around the North Celestial Pole, that around the South Celestial Pole is devoid of bright stars. The constellation Hydrus (the Lesser Water Snake) could not look any less like a snake, the group comprising the triangle formed from Alpha, Beta and Gamma Hydri, an unspectacular trio of 3rd magnitude stars. Although somewhat obscure, Hydrus can be picked up fairly easily by using the bright star Achernar (Alpha Eridani) as a guide; Achernar lies within a degree or so of Alpha Hydri.

Adjoining Hydrus is the similarly obscure constellation Tucana (the Toucan), an irregularly shaped circlet of stars which does, however, contain several deep-sky objects of interest to the backyard observer (see chart *right*). One of these is Beta Tucanae, a beautiful triple star. Beta$_1$ and Beta$_2$ form a double star with yellowish components of magnitudes 4·4 and 4·8, situated 27·1" apart. 5th magnitude Beta$_3$ lies close by. The Beta$_{1,2}$ system is resolvable from Beta$_3$ with the naked eye provided the sky is really dark and clear, and presents a lovely sight through binoculars. All three stars are double again, although their components are all too faint and close to their primaries to be resolvable without large telescopes. Splendid though the Beta Tucanae system is, however, Tucana's main objects of interest are the Small Magellanic Cloud and the two globular clusters NGC 362 and 47 Tucanae, for which detailed descriptions follow.

The South Celestial Pole (SCP) lies within the constellation Octans (the Octant), the brightest star of which is Nu Octantis, which glows at a feeble magnitude 3·74! The southern equivalent of Polaris (the northern Pole Star) is Sigma Octantis which is located within 1° of the SCP. Sigma is barely visible to the unaided eye, shining at magnitude 5·46, and cannot be seen without optical aid unless the sky is really dark and clear. To locate Sigma, use the accompanying chart (see *right*) to pick out the main shape of Octans and you should be able to locate Chi and then Sigma Octantis, the pair of stars emerging from the line between Beta and Delta Octantis. Patience may be required, so do persist in your search!

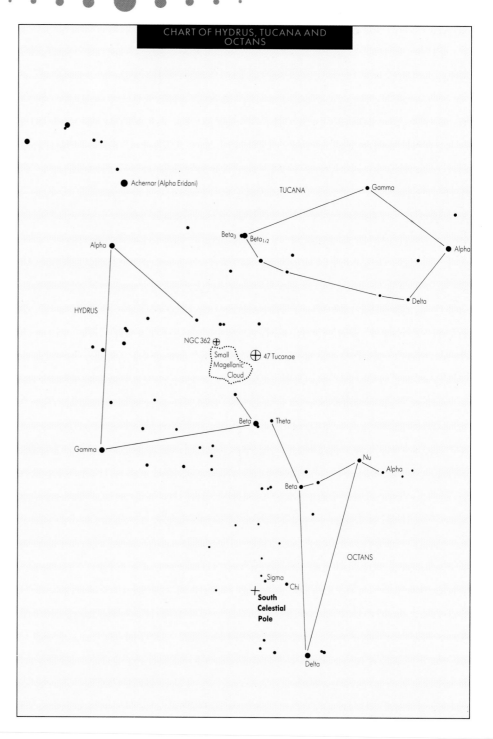

CHART OF HYDRUS, TUCANA AND OCTANS

THE SMALL MAGELLANIC CLOUD

The Small Magellanic Cloud (SMC) is an irregular galaxy and a member of the Local Group of galaxies (see Galaxies, pages 116-117). Located at a distance of 200,000 light years, the SMC lies across the border between Hydrus and Tucana. It is easily seen with the naked eye as a hazy, elongated patch of light. Sweeping with a telescope or good binoculars will reveal a number of prominent nebulae and star clusters scattered across the SMC (see *right*).

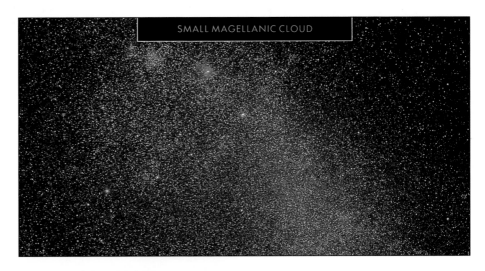

SMALL MAGELLANIC CLOUD

47 TUCANAE

GLOBULAR CLUSTER 47 TUCANAE

Tucana also contains a fine pair of globular clusters one of which, 47 Tucanae, is seen here. 47 Tucanae (NGC 204) is one of the finest globulars in the sky, second only to Omega Centauri (see Star Clusters, pages 124-125) in terms of visual impact. It is easily visible to the naked eye as a hazy, starlike object shining with an overall magnitude of around 4·5.

47 Tucanae is a huge system with a diameter in excess of 200 light years and a total actual luminosity of over a quarter of a million times that of the Sun. Telescopes of 100-mm (4-in) aperture and upwards will resolve individual stars at the cluster's edge, the degree of resolution increasing rapidly as larger instruments are turned towards the cluster (see photograph *left*).

The other globular in Tucana, NGC362, is much less condensed and therefore fainter than 47 Tucanae. It is only dimly visible to the naked eye, and binoculars may be needed in order to see it against the backdrop of starfields at the edge of the Small Magellanic Cloud. As with 47 Tucanae, telescopes of 100-mm (4-in) aperture will start to resolve individual cluster stars. The proximity of Tucana's two globular clusters to the SMC is merely the result of a line-of-sight effect. 47 Tucanae lies at a distance of 16,000 light years. The distance of NGC 362 is some 30,000 light years.

DORADO AND MENSA

The constellations Dorado (the Goldfish) and Mensa (the Table Mountain) are notable chiefly for the presence of the Large Magellanic Cloud that straddles their borders (see chart *below*).

Beta Doradus is a Cepheid variable whose magnitude varies between 3·8 and 4·7 over a period of 9·84235 days. Delta Doradus (magnitude 4·34), Zeta Doradus (4·71), Eta₂ Doradus (4·88) and Lambda Doradus (5·13) are useful comparison stars.

Mensa is one of the smallest constellations in the entire sky. It is also the dimmest, its leading stars Alpha and Gamma shining at magnitudes 5·09 and 5·19 respectively! However, it can be traced out either with the naked eye (under clear skies) or with binoculars, by following its path away from the southern borders of the Large Magellanic Cloud.

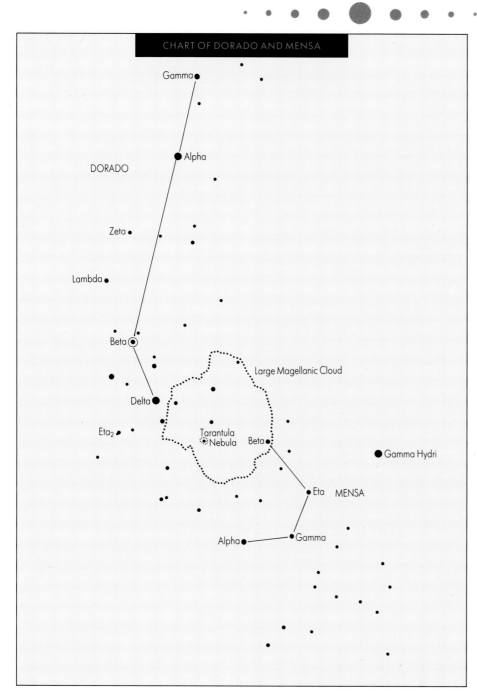

CHART OF DORADO AND MENSA

THE TARANTULA NEBULA

The Tarantula Nebula (NGC 2070) is the largest-known diffuse nebula in the Universe. Its outer filaments and streamers spread out to cover an area of space some 1,750 light years across and the fact that it can be seen with the naked eye at a distance of 190,000 light years bears testimony to its colossal size and brilliance. Its diameter is almost double that of the huge NGC 604 nebula in M33 (see Galaxies, pages 116-117) and its mass is equal to around 500,000 times that of the Sun. The region around the Tarantula Nebula can be seen in this photograph (*below*).

REGION AROUND TARANTULA NEBULA

THE LARGE
MAGELLANIC CLOUD

The Large Magellanic Cloud (LMC), like its counterpart the Small Magellanic Cloud, is a member of the Local Group of galaxies. Situated at a distance of around 190,000 light years, the LMC is plainly visible to the unaided eye, and repays time spent sweeping it with bin-

WITHIN THE LMC

The LMC contains a wealth of deep-sky objects for the backyard astronomer. Careful examination of this photograph will reveal a host of stars, star clusters and nebulae scattered across both the face of the LMC and around its borders. Particularly prominent near the south-eastern edge of the LMC (upper left in this picture *below*) is the Tarantula Nebula (see opposite page). Around 700 open clusters have been found in the LMC together with over 60 globulars and several hundred nebulae, including 400 planetary nebulae. A large number of these objects is visible through moderate telescopes, and the backyard astronomer could easily spend many hours gazing at the many and glorious sights the LMC has to offer.

oculars or a wide-field telescope.

The LMC, captured in this stunning wide-angle photograph (*below right*), is visible even in moonlit skies and has been likened to a detached portion of the Milky Way. A sizeable system with a diameter of 50,000 light years, it is situ-

ated some 22° from the SMC, corresponding to an actual separation of around 80,000 light years. The total mass of the LMC is thought to be roughly 10 per cent of that of our own Milky Way Galaxy while its combined luminosity equal to 2,000 million Suns.

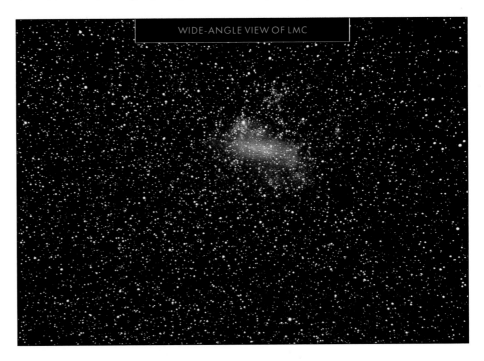

WIDE-ANGLE VIEW OF LMC

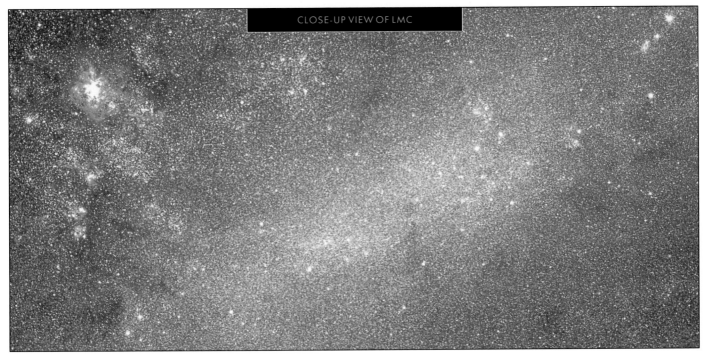

CLOSE-UP VIEW OF LMC

STAR CLUSTERS

Our Galaxy plays host to an enormous number of star clusters and associations which contain anything from a dozen or so to many thousands of individual stars. These stellar gatherings are divided into two categories: open clusters and globular clusters. Open clusters are found within the rich starfields of the main galactic plane, while globular clusters are seen outside the main plane of the galaxy. As we shall see, each type of cluster has its own individual characteristics which clearly sets it apart from the other.

THE DISTRIBUTION OF STAR CLUSTERS

Many globular clusters are located in the galactic halo (see diagram A), a spherical volume of space surrounding the galaxy. Those within this region orbit the galactic centre in eccentric and sometimes highly inclined orbits. Some globulars are located in the galactic plane. Open clusters are found within the galactic disc (see diagram B) and orbit the galactic centre in circular paths.

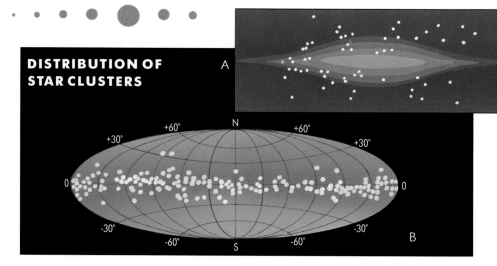

DISTRIBUTION OF STAR CLUSTERS

GLOBULAR CLUSTER M92

This is one of the most visually stunning of the 150 or so globulars known. Lying at a distance of around 35,000 light years, the total light output of M92 is equal to around 250,000 times that of the Sun. There are few globulars to match M92 as a telescopic showpiece, although one of these, M13, also lies in Hercules.

Also known as the Great Hercules Cluster, M13 quite often steals the limelight from neighbouring M92.

As with other globular clusters, the stars within M92 are relatively tightly packed (see *below*). As our gaze travels down towards the central regions of this huge stellar gathering, the stars are seen to become more and more compact. This leads to the combined gravitational attraction, which helps globulars retain their characteristic spherical shapes.

OPEN CLUSTER M67

Open clusters are much more common than the globulars, and around a thousand have been observed and catalogued. They can contain anything from a dozen to several hundred stars. Like the Hyades and Pleiades, M67 (see *below*) is a rich cluster with around 500 member stars. Also known as the Beehive, M67 shines from a distance of around 2,500 light years. It is one of the oldest-known open clusters, its estimated age around 10,000 million years. Indeed, the ancient stars within this cluster are more like those found in globulars.

GLOBULAR CLUSTER M92

OPEN CLUSTER M67

OMEGA CENTAURI

Clearly visible to the naked eye, Omega Centauri is probably the best example of a globular cluster. This magnificent object shines at around 4th magnitude and is the brightest of the globulars. It was first noted over 1,800 years ago by the Greek astronomer Claudius Ptolemaeus, and was catalogued as a star by the German astronomer Johann Bayer in the early 17th century, but its true nature only became apparent when the English astronomer Edmund Halley observed it in 1677. The diameters of globular clusters can range from 20 to 30 light years up to several hundred light years. Omega Centauri itself has a diameter of 350 light years and lies some 17,000 light years away. It is one of the richest globulars known, and is thought to contain in excess of a million stars. As with other globulars, the stars in Omega Centauri are of the older type usually associated with galactic nuclei.

Well over 150 variable stars have been spotted in Omega Centauri, the majority of which are of the RR Lyrae type. Other variables discovered in Omega Centauri include several Cepheids and a number of long-period variables. (A description of the different types and characteristics of variable stars can be found on pages 114-115). Omega Centauri's member

stars become more tightly packed towards its central regions. The average distance between the stars at the cluster centre is believed to be only around a tenth of a light year or so. Rotation of Omega Centauri is thought to have caused the cluster's slightly elliptical shape, the effect of which is noticeable in this photograph (see *below*).

GLOBULAR CLUSTER OMEGA CENTAURI

HYADES & PLEIADES

Many open clusters are visible to the unaided eye including the two examples seen here. The Pleiades, the compact cluster visible just above centre of picture (see *below*), is made up of between 250 and 500 young, hot stars, typical of those found throughout the spiral arms of our own and other galaxies. As with many open clusters, the space between the

stars in the Pleiades is seen on photographs to contain traces of dust and gas, remnants of the original nebulae from which these objects were formed. In comparison, globular clusters contain little if any such interstellar material. As we have seen (see pages 82-83), the stars in the Pleiades are thought to have formed between 60 and 70 million years

or so ago, making this cluster much younger than the neighbouring Hyades, located a little way to the south-east (the V-shaped cluster below centre of photograph). The Hyades is believed to be some 800 million years old, therefore containing much older stars. As a result of internal gravity-induced motions within the cluster, the stars in the Hyades are more spread out than those in the Pleiades. Generally speaking, the older a cluster gets, the more its member stars dissipate into surrounding space.

As with open clusters in general, both the Hyades and Pleiades are irregularly shaped, unlike their much more uniform counterparts, the globulars. The Hyades is one of the closest open clusters to us, lying at a distance of around 145 light years. Its proximity ensures that many of the Hyades stars, a large number of which are much fainter than the Sun, are within the light grasp of our telescopes. Open clusters can range in size up to several tens of light years, the diameter of the Hyades being roughly 75 light years.

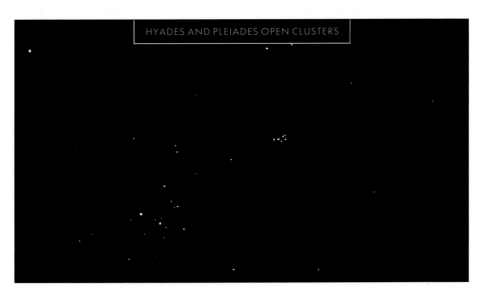

HYADES AND PLEIADES OPEN CLUSTERS

Page numbers in *italics* refer to illustrations.
Entries accompanied by (lunar) indicate features on the Moon